PENGUIN BOOKS
IQ, EQ, DQ

Dr Yuhyun Park is the founder of the DQ Institute and a world-leading expert in digital skills and child online safety. She created the Digital Intelligence (DQ) concept and framework, which was officially approved as the world's first global standard related to digital literacy, digital skills, and digital readiness by the IEEE Standards Board (IEEE 3527.1™). It was originally endorsed by the OECD, IEEE Standards Association, and World Economic Forum in 2018 as the foundation to build global standards and a common language for digital literacy and skills. Dr Park also developed the Child Online Safety Index, which is the world's first real-time metric tracker to help nations better understand their children's online safety status, and leads the #DQEveryChild initiative, which is a global digital citizenship movement that has empowered children, families, and teachers in more than 80 countries to date. In addition, she serves in various global leadership positions related to digital economy and education, including as the International Lead for Digital Economy in the G20 Civil Society (2020), head of the EQUALS ITU Digital Skills Coalition, and founder of the Coalition for Digital Intelligence. She has received numerous international awards, including recognition as a World Economic Forum Young Global Leader, Ashoka Fellow, Eisenhower Fellow, and multiple UNESCO prizes. She co-authored the *Dictionary for Economics*, which is the most widely used dictionary for economics in Korea. Her academic experiences include serving as an adjunct professor at Yonsei University in Korea and as a director at Nanyang Technological University in Singapore. Dr Park completed her Ph.D. degree and post-doctoral studies in biostatistics at Harvard University.

This book is about the safe and moral use of digital technology and reminds us that young minds may be especially at risk until they develop adequate digital intelligence (DQ). A must-read if you care about digital guardrails in our society.
—Vint Cerf, Internet Pioneer, One of the fathers of the Internet, and recipient of the US National Medal of Technology

Yuhyun Park is one of the world's most successful social entrepreneurs. She has enabled many and very different societies to bring millions of their kids digital understanding and independence—and therefore far healthier, more successful lives. Yuhyun's success here reflects her deep understanding of the digital revolution and the urgency of humans guiding it for good and not becoming its dependent pawns. In this book she shares that understanding with rare clarity—a gift.
—Bill Drayton, Founder, Ashoka: Innovators for the Public, and inventor of the terms 'social entrepreneur' and 'changemaker'

A must read story of passion and inspiration that has become the language of the 21st century; a language we need to ensure we are all able to embrace.
—Sir Peter Estlin, London

After more than 30 years dedicated to helping people live fuller, richer, easier lives with technology, it fascinates me to see that the more we evolve, the more we speak of technology in de-humanized terms. The tool becomes de-coupled from the user and the usage case, the means becomes an end in and or itself, and the end comes back to haunt the society because we have failed to establish the foundations on which technologies of the 4th industrial revolution should have risen.

What Dr Park does with her work on DQ is extremely valuable because it takes us back to foundations, and gives us a chance to catch up where we have lagged behind, especially with regards to the most vulnerable parts of the society: Our children.

With #DQEveryChild, we now have a globally acceptable framework to support the growth of our children, not as victims of technology, but as fully aware digital citizens, responsible co-creators and visionary leaders of the digital world. This book takes us through her journey, which I am privileged to have been a part of, and into the depths of various dimensions of children's development in the digital sphere—an invaluable resource for not only for educators, parents and but also for each and every one of us in the technology space, if we are to ground ourselves as contributors to the well-being of the society and not as a factor in its disintegration.

—Kaan Terzioglu, Group Co-CEO, VEON

I can tell you that the mission that Dr Park has embarked on and made strides in is of immense importance to the way we navigate the future. Her contribution in ensuring learners in education systems around the globe can flourish in an increasingly digital world is profound and has far reaching impact. To say she is a visionary is a massive understatement and I am certain that when future generations look back to this period in time they will refer to her in the same way as we do, today, of the icons who built the internet.

—Vikas Pota, Founder, T4 Education

We are living in the middle of one of the most profound transformations in human history as the digital world becomes an integral part of our daily lives. Yuhyun's story poignantly captures this digital transformation. The evolution of the digital intelligence (DQ) concept as Yuhyun journeys from a concerned mother to a trailblazing social entrepreneur and researcher is what makes DQ so authentic and timely. DQ captures the human imagination and paints a path forward to not only address some of the greatest challenges of our time but to also imagine what is possible for the future of education in the digital era.

—Bertil Andersson, the third President of
Nanyang Technological University,
the former board member of the Nobel Foundation

The world is rapidly changing with the ongoing digital transformation, and a large fraction of human activity, including education, is already taking place in the digital world. Nowadays, almost 1 billion children are online and, by 2050, all of them will be. We must pay greater attention to the possible shadows of the digital world and that is why I believe digital intelligence (DQ) is so critical to help children thrive in the digital world.

—Doh-Yeon Kim, Chairperson of Ulsan Education
Foundation, Former Minister of Education,
Science & Technology , Republic of Korea

IQ, EQ, DQ

New Intelligence in the AI Age

YUHYUN PARK, PhD

PENGUIN BOOKS
An imprint of Penguin Random House

PENGUIN BOOKS

USA | Canada | UK | Ireland | Australia
New Zealand | India | South Africa | China | Southeast Asia

Penguin Books is part of the Penguin Random House group of companies
whose addresses can be found at global.penguinrandomhouse.com

Published by Penguin Random House SEA Pte Ltd
9, Changi South Street 3, Level 08-01,
Singapore 486361

First published in Penguin Books by Penguin Random House SEA 2021

ISBN 9789814954396

Typeset in Adobe Garamond Pro by Manipal Technologies Limited, Manipal

www.penguin.sg

To Isaac and Kate

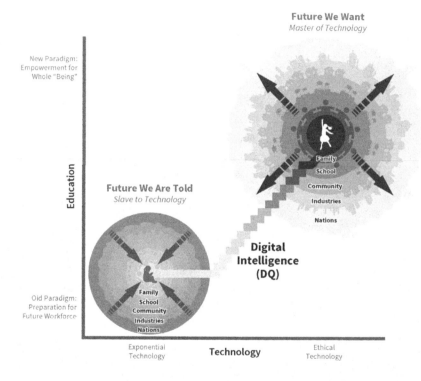

Future We Want
Master of Technology

New Paradigm:
Empowerment for
Whole "Being"

Education

Future We Are Told
Slave to Technology

Family
School
Community
Industries
Nations

Family
School
Community
Industries
Nations

**Digital
Intelligence
(DQ)**

Old Paradigm:
Preparation for
Future Workforce

Exponential
Technology

Technology

Ethical
Technology

Schematics of the book

Contents

Reflection

In December 2008, there was a horrible incident in Korea, called 'Nayoung's case'[1]. An eight-year-old girl, Nayoung (fake name), was kidnapped, brutally raped and beaten to near-death by a paedophile, Mr Cho Du-Soon, at 8 a.m. one morning, when she was on her way to school. The whole country was outraged by this cruel act. What shocked people most was the sheer ordinariness of the offender, Mr Cho. He was an average middle-aged man, who lived in the same neighbourhood as Nayoung's family. Later, the police found gigabytes of child pornography videos on Mr Cho's personal computer.

At the time, I was pregnant, living in Palo Alto in the US. Just like any other mom, I was extremely upset when I encountered this news. I attentively followed the story through online news portals.

One day, I saw a drawing image on one of the major news sites in Korea. It was drawn by Nayoung on her hospital bed, expressing her wish that the court would give Mr Cho the maximum punishment. While scrolling down this news article, I saw another image below Nayoung's drawing. It was a photo advertisement of a naked girl on her bed with the large print, 'A photo gallery of an innocent 16 year-old-girl who invites you to her bedroom'.

I could not breathe. I felt a sharp pain in my tummy. Uncontrollable tears flowed from my eyes. I did not know why. It was likely due to the hormonal emotional roller-coaster that is pregnancy. A huge wave of complex feelings in varying combinations of sadness, anger and guilt that I could not process, surged through me. I found myself keep repeating, 'I'm sorry, I'm sorry, I'm sorry'. I'm sorry to whom? To Nayoung, to Nayoung's parents, or to whom?

Back then, I was working at the Boston Consulting Group as a senior analyst and consultant, specializing in technology and the digital media industry. The 2000s were tough for traditional media companies, due to the rapid rise of Internet businesses. People stopped reading printed newspapers and watching shows on the TV—newspaper circulation went down dramatically and traditional TV broadcasting could not gather enough viewership. That meant they no longer captured enough eyeballs of people, and companies became less interested in putting their ads on traditional media outlets—newspapers or TV, in other words, their traditional way of getting ad revenue did not work any longer. Blockbuster went bankrupt in 2010, which symbolized the collapse of the traditional media industry.[2]

Since the early 2000s, the Korean government had invested heavily in the Internet and Information and Communication Technology (ICT) industries as part of a national digital transformation agenda to escape the Korean Financial Crisis. While Korea was rising to become one of the most innovative and wired countries in the world, the traditional media companies in Korea had to desperately seek out a new business model. Clickbaits with lewd and provocative content that brought quick and easy cash from online advertisement, were one of the new revenue sources for these companies.

The online news page hosting the image of Nayoung's drawing and the child porn ads together, was not coincidental. In fact, the moment I saw that news website put Nayoung's drawing and an under-age girls' semi-porn site, I realized that the victim was not just one eight-year-old girl in Korea. The offender was also not merely one child porn addict. I concluded that my children, your children, and all children around the world were at risk. And hence, we all are responsible.

We are so quick to blindly praise new technology. We praise anonymity and freedom of speech online. We praise Internet start-up billionaires. But we neglect our children, who find themselves in the blind spot of a dark corner of the Internet. We neglected 'Infollution'—information pollution or the negative consequences of technological advancement—such as cyberbullying, technology addiction, privacy invasion, child-porn, online grooming and more. While pollution has harmed our earthly environment, Infollution has polluted our and our children's minds. We have neglected our duty to carefully consider the impact of Infollution, especially on the weak and young.

In September 2009, I went back to Korea and decided to start a not-for-profit initiative called 'InfollutionZERO' (iZ) with the vision of making Infollution around our children equal to ZERO. Today, in 2020, many parents are concerned about their children's screen time and exposure to cyber risks. But in 2009, unfortunately, not many parents were aware of or cared about their child's safety online. And many industry and government leaders did not welcome such initiatives as they came across as making unnecessary complaints against a major national agenda of digital transformation.

An even more unfortunate thing was that I was not smart enough to fully understand this reality. It did not take long for me to find out that people thought of me as an idealistic weirdo and/or a total loser. When I officially opened the office in January 2010, a few major newspapers in Korea wanted to interview me, not because of the InfollutionZERO agenda, but because of my elite profile. They all asked me 'why did you volunteer to become a secondary citizen after all of your academic study and hard training in international businesses?' At the time in Korea, starting a not-for-profit start-up (especially to address an issue that 'nobody cared about') was considered a failure, unless you had a great mission that all acknowledged, or you had a political ambition, or a personal tragic story to struggle against. I had none. I just saw the problem and I thought I could work on it. That was all. Nothing heroic, nor political, nor vindictive.

A few months later, I was asked to meet one of the officials of the Blue House, who saw one of my newspaper interviews. He asked me, 'what are you trying to achieve?' I immediately sensed that he also wanted to check if I had any political ambition or hidden agenda.

I told him, 'within ten years, I will set the global standards for digital safety and empowerment for children. So that every ICT company and policymaker around the world can no longer ignore the need to make child protection online and digital citizenship education, a top priority.' He paused for a moment in surprise and then gave me a look. I could read his eyes silently dismissing me, 'who do you think you are?' He smiled and said, 'Well, I don't think it will be possible, but good luck.'

Really?

Honestly, I did not have such a grand ten-year vision. In fact, I had never thought about it at all. I have struggled to survive one day at a time since I started InfollutionZERO. Somehow, it just fell out of my mouth. Maybe I said it out of frustration and anger, as people did not care seriously about my cause. Only a few people truly supported it. At almost every meeting regarding fundraising or activism, I was either ignored or laughed at.

However, that earlier conversation gave me a new vision and goal. 'Why don't I set up the global standards for safety on the Internet within ten years?' On that day, I wrote in my diary that I would work on this social impact journey for the next ten years and would develop the following three things: a global standards framework that every nation can adopt; a global child education programme that every child can use, and a global index that every nation shall pay attention to. Since then, I moved to Singapore and continued working on child online safety and digital education. I developed the concept and framework of DQ (Digital Intelligence) and many other related programmes, and set up the DQ Institute in association with the World Economic Forum in 2017. Fast forward . . .

At 5 p.m. on 26 September 2018, I was sitting in a press conference room of the World Economic Forum in the UN General Assembly Week in New York City, together with Gabriela Ramos, the OECD Chief of Staff and Sherpa to the G20; Karen McCabe, a Senior Director of the IEEE Standards Association, and Eric White, who was leading the 'Internet for All' programme for the World Economic Forum. These three international organizations and the DQ Institute announced the launch of the Coalition for Digital Intelligence (CDI)[3], a global coalition to promote digital intelligence (DQ) around the world. All three international organizations agreed to use the *DQ Framework* that my team and I developed, as a global standard for digital literacy, skills, and readiness. The child online safety was a core component of the *DQ Framework*.

While speaking at the stage of the press room, I remembered myself in 2010, at the Blue House, telling an officer about my ten-year goal, which was completely nonsensical back then. I realized that it was my eighth year since the day I told him about my vision. Chilled. Overwhelmed. Afraid.

Because I did not work towards setting the standards, nor plan anything strategically to be prepared for that stage, I did not have the luxury to think about that vision or my three goals. I tried to seize every opportunity and work on anything that I could do, which was related to child online safety and digital citizenship education. I worked with children to develop children's digital citizenship programmes. I conducted training for teachers and parents. I ran public campaigns and social outreach programmes with the ICT and media companies. I worked on policies and regulations for governments. I conducted academic research. I supported international organizations like ITU, G20, UNESCO, UNICEF, UN, OECD, and more, in all sorts of capacities. I made so many mistakes. I faced many failures. But I also had some successes.

From March 2017, my team and I started running #DQEveryChild,[4] a global digital citizenship movement with the aim of empower every child with responsible digital citizenship, using the DQ World[5] e-learning programme for eight to twelve-year-olds. Within three years, this digital citizenship education reached more than 1 million children in over eighty countries, via both online and offline channels. Thankfully, over 100 partners, including ICT companies like Singtel and Twitter, government agencies like Singapore IMDA (Infocomm Media Development Authority) and Korea IFEZ (Incheon Free Economic Zone), international organizations like the World Economic Forum and UNCEF, and many civil organizations like JA Worldwide and TOUCH Cyberwellenss, worked together with me. And in February 2020, based on the past three years of #DQEveryChild, we published the *Child Online Safety Index* (*COSI*), the world's first real-time measure to help nations better understand their children's online safety status. In June 2020, I made an official suggestion to the G20 Digital Economy Task Force[6] that *COSI* should be included in the G20 Measurement of Digital Economy. Finally, in 24th September 2020, I was told by the IEEE Standards Association—which is one of the world's biggest technological associations and authorities that defines global standards across sectors and related to various technologies, that Digital Intelligence (DQ) was officially approved as the global standard related to digital literacy, digital skills, and digital readiness (IEEE 3527.1™ Standard)[7]. It was exactly the tenth year since I had stated my ten-year vision in 2010.

On 10 October 2019, the inaugural international 'DQ Day'[8] was launched in two major cities: New York and London. In New York City, we launched the DQ Day conference at the World Economic Forum, together with IEEE, IBM, World Economic Forum, and many other partners. In London, the City of London launched a nationwide digital skills initiative in collaboration with BT, Accenture and other companies, with the title Future.now. It was a day of true celebration for me. At the DQ Day conference, I thanked all of the partners and our team, and celebrated the fact that our initial three ten-year goals have been materialized through the *DQ Framework* (setting a global standard framework that every nation can adopt), DQ World (a global children's education programme that every child can use), and *Child Online Safety Index* (a global index that every nation shall pay attention to). I know that it was never 'I' who achieved this. It was a collective 'we' who achieved it together.

On that day, I was asked by many people, 'so now, what are you planning to achieve in the next ten years?' Thankfully, this time, it was asked with genuine interest and sparkling passion.

This book talks about digital skills, future education, technology, and ethics—the buzzwords even more so, after the COVID-19 pandemic. I do not dare to compose this book as the most intelligent and intellectually challenging book to read on these topics. Rather, I want to tell you the stories that I have personally experienced and the people who have inspired me and worked together with me in the past ten years. Especially, I want to give you the contrasts between two different views that I encountered on each topic—they are not necessarily the 'right' or 'wrong', rather they are two different views, but on a spectrum that moves from one end to the other. I was trained as a mathematical statistician. My worldview as a researcher is rather grey with a probabilistic scale; seldom in a black-and-white mode. When people asks me about my opinion on a certain matter, as a researcher, I tend to say that it can be likely correct, but that it could also be wrong in a certain context. However, my worldview as a social impact leader as well as an activist can be quite black and white. Why would it not be? That is the beauty of being a social impact leader, rather than a pure university scholar. I will leave it to you to decide on your stand on those two views.

In my favourite movie, *Great Expectations*,[9] Ethan Hawke started the movie with the quote, 'The colour in memory depends on the day. I won't tell the story the way it happened. I'll tell it the way I remember it.' I am writing this book in that same manner. People, places and the feelings that I remember at the moment of writing, will deliver the story. So this book is strictly written from my perspective, with the help of insights that I have learned in the last ten years of my journey. So, I do not expect all of you to agree with what I believe and how I see the world. Anyhow, I wish this book can offer some different perspectives to you on how we can create a better world for our children. So I hope we can dream together. Who knows where we will be, in ten years?

27 October 2020
Yuhyun Park

DQ Global Standards

Setting the DQ Global Standards
is not about championing the world.

It is about *drawing the line.*

So that not a single child falls below this line.

Above this line, every child
will *walk safely, run confidently* and *fly boldly*
in the AI age.

1

Fundamentals

Before we discuss modern education and technology, I want to discuss what the fundamental assumptions and beliefs in our society are. I encourage you to ask yourself the following five questions. I believe without having your own answer to these questions, we can't have a meaningful discussion on education and technology in this age.

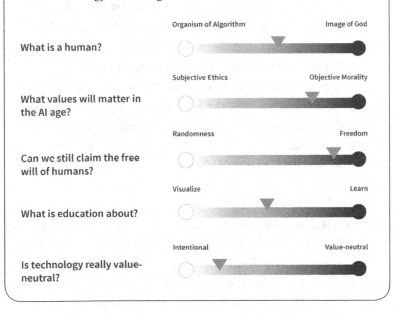

Human: Image of God vs. Organism of Algorithm

All are true and all mysteriously coinhere in that one person.

—*John Polkinghorne*

Look at yourself in a mirror. What do you see?

You will see your body, for sure. And you already know that you have your own mind for thinking, emotion, feeling and will, even though you don't see it. Is that all that you have? Your body and mind? Or do you find more than that within yourself?

I am talking about the philosophical or religious stuff. I am a boring and dry statistician, but even I believe that 'What is human?' is the most important question that we need to ask ourselves today, before we talk about modern technology or education.

In March 2015, I received an email from my boss at the university, 'Congratulations, you have been selected as a 2015 Young Global Leader (YGL) at the World Economic Forum (WEF)!' What? I was not 'young' anymore, and I'm certainly not a 'global leader'. I sent an email to the WEF and asked, 'Is the person that I see on your website, really me?' They said yes. Wow. How flattering this title was.

In the same year, I went to my first YGL summit at Geneva and met Professor Klaus Schwab, the founder and chairman of the WEF. At that time, he coined the term and devised the concept of the 'Fourth Industrial Revolution (4IR)'—he introduced the concept at that meeting before he published his book, *The Fourth Industrial Revolution*,[1] in the following year. I was genuinely impressed with the concise and clear paradigm of the 4IR. He said, 'this fourth industrial revolution is unlike anything humankind has previously experienced. New technologies are merging the physical, digital and biological worlds in ways that will create both, huge promise and potential peril.' He ended his speech with a question, 'What is human? What defines us as human? We may need to redefine our humanness and values in the 4IR.'

Yuval Noah Harari, a rock star historian and author of *Sapiens*[2] and *Homo Deus*[3], described the human being as an organism which is composed of algorithms. He described how even our feelings and emotions are the products of biochemical algorithms and our decisions, thus, can be predicted, based on probability and optimization. He warned of the soon-to-be end of humanism which stemmed from the belief that I and my own authority—my feelings, desires and free will are governed by 'me'—the highest authority of all. Our body, mind and decisions can, however, be hacked when there are enough data and computing powers to understand the algorithms of a human. In other words, The AI (Artificial Intelligence) and biotechnology will soon be advanced enough to unlock this algorithm of human—the black box that is a human being.

Such a (cold) view on humans has been agreed upon by some scientists and supported by technological breakthroughs. Among whom, I found the research of Craig Venter, a world-leading biologist who was most famous for his role in mapping the human genome to decode the origin of life, quite daring and bold. In 2010, he and his team made the first manmade self-replicating synthetic organism, a manufactured version of *M. mycoides*' genome, which was transplanted into a different *Mycoplasma* species.[4] I met Craig Venter in 2012, when I organized the Molecular Frontier Symposium at Singapore. He introduced in the Symposium, this concept of 'synthetic life'—an organism whose parent is not a living being, but a bunch of digital code. This, again, begs the question: 'then what is life and what is human?'

If we are only an organism having a body and a mind, our 'human question' can be reduced down to mere bio-chemical-computational equations. And we don't even need to question ourselves about what human is. We can consider ourselves an advanced animal, having a more complex bio-chemical composition and more dynamic algorithms. Human, then, can be made upgradable with genetic engineering, or even replaceable by a healthier version of a cloned self.

However, the reason that we—even top scientists, historians, and global leaders—are repeatedly asking themselves what human is, as technology advances, maybe because deep down in our hearts, we see ourselves as more than a mere organism, composed of a body and a mind. In Korean, the word for 'human' is 'In-Kan (人間)', whose direct translation is 'between people'. In Eastern culture, the meaning of 'human' lays emphasis not solely on the individual living organism, but on the community and the relationships between individuals. Values, society's laws and social identity are all considered a crucial part of how 'human' is defined. In case of such an understanding, we see ourselves as a spiritual and communal being that has a body and a mind. Human is considered the image of God—we see that we have an intrinsic and ultimate worth as a human, the image of God.

We see that image of God when people uphold values higher than themselves; when people seek life's purpose and meaning; and when people pursue social justice and righteousness. We see greatness when we hear about the stories of people like Mahatma Gandhi or Nelson Mandela, who fought against violence and power, with non-violence and peace. And we humble ourselves before the higher law of Love, when we witness people like Mother Teresa, who sacrificed themselves for people in need.

These two views don't need to be mutually exclusive. Both scientific understanding about our bodies and mind, and philosophical and religious beliefs about our spirits, can be considered as different facets of the one truth about 'what is human'.

I love how Dr John Polkinghorne, a former professor of mathematical physics at University of Cambridge who later became an Anglican priest, answers this question in his book, *One World: The Interaction of Science and Theology*.[5] 'Reality is a multi-layered unity. I can perceive another person as an aggregation of atoms, an open biochemical system in interaction with the environment, a specimen of homo sapiens, an object of beauty, someone whose needs deserve my respect and compassion, a brother for whom Christ died. All are true and all mysteriously coinhere in that one person. To deny one of these levels is to diminish both that person and myself, the perceiver; to do less than justice to the richness of reality.'

What is your view? What is human?

Values: Objective Morality vs. Subjective Ethics

Every human being has intrinsic worth.

In the 2019 Davos World Economic Forum Annual Meeting, the first official meeting of the Coalition for Digital Intelligence[6] was held to discuss how to coordinate various global efforts for digital skilling and promoting Digital Intelligence (DQ) effectively. I was asked to prepare the one-pager for the meeting in order to help the business and political leaders understand the *DQ Framework*, which was identified to serve as global standards for digital literacy, skills and readiness. So I sent the image below.

At the meeting, I found that the printed image was different. The bottom red line of 'Universal Moral Values' was removed. The person in charge of the meeting in the WEF, read my face and told me, 'Yuhyun, please don't get upset. I cut this bottom line without your consent as I don't want people to start the 'value' discussion. As we have a short meeting time, let us keep people's focus on digital skills.'

I have also experienced on several occasions, when someone talked about importance of 'value' in the 4IR, people immediately started arguing, 'whose value are you talking about? Western or Eastern values? Is it Judeo-Christian value? What about LGBT value?' I was told that the Davos is a neutral platform that should not be biased toward any particular values, related to certain religions, culture or countries. So I understood his concerns and I didn't say anything about his decision to remove the 'Universal Moral Value' line at that meeting.

IQ EQ DQ

Global Standard Framework for Digital Literacy, Skills and Readiness

Digital Intelligence (DQ) is a comprehensive set of technical, cognitive, meta-cognitive and socio-emotional competencies that enable individuals to face the challenges of and adapt to the demands of digital life.

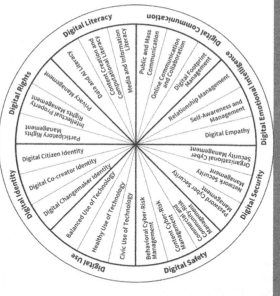

Digital Literacy
- Content Creation and Computational Literacy
- Data and AI Literacy
- Media and Information Literacy

Digital Communication
- Public and Mass Communication
- Online Communication and Collaboration
- Digital Footprint Management

Digital Rights
- Privacy Management
- Intellectual Property Rights Management
- Participatory Rights Management

Digital Emotional Intelligence
- Relationship Management
- Self-Awareness and Management
- Digital Empathy

Digital Identity
- Digital Citizen Identity
- Digital Co-creator Identity
- Digital Changemaker Identity

Digital Security
- Organizational Cyber Security Management
- Network Security Management
- Password Security Management

Digital Use
- Balanced Use of Technology
- Healthy Use of Technology
- Civic Use of Technology

Digital Safety
- Behavioral Cyber-Risk Management
- Content Cyber-Risk Management
- Commercial and Community Risk Management

Digital Citizenship

Ethically use technology

Digital Creativity

Co-create new values

Digital Competitiveness

Solve global challenges for humanity

Universal Moral Values

However, unfortunately, we can't avoid this discussion of 'morality' and 'values' if we want to talk about technology and education in today's world. Rather, human value needs to be discussed more widely as well as more intensely, compared to any other time in history. Because the value discussion can no longer stay within territory of religion or philosophy in the 4IR. It is a very practical question for technologists, scientists, and engineers to think through what and how technology must be developed and designed. And it is a practical question for policymakers and business leaders to decide how and what technology needs to be deployed and regulated.

Let's think about a widely used ethical dilemma. Let's imagine that you are driving a car with a broken brake. You need to decide to turn the wheel either left or right, or you will die. If you turn it to the left, you will kill a mom and her baby, and if you turn it right, you will kill a group of seniors. If you keep the wheel straight, you will kill yourself. What will be your moral choice? If it is up to your decision, you will need to take the responsibility of the consequent deaths. But if you are riding a self-driving car that has a deterministic or probabilistic algorithm to respond to such a situation, who will be responsible for the consequent deaths from the algorithm's decision? Moral values of the engineers who designed the car algorithm, those of business leaders who put these cars in the market, and those of policymakers who set related regulations and policy, are no longer in the realm of personal philosophical or religious matters. They practically decide the destiny of millions of lives.

Moreover, every citizen's moral values are also reflected in machines. Our everyday small decisions on what we say, how we behave, and what we click online, are reflected in the next generation AI algorithms and potentially shape our society. A good lesson can be found in the failure of Microsoft AI chatbot Tay[7] in 2016. Microsoft launched its machine learning chatbot, named Tay, and encouraged users to talk to Tay more to make him/her smarter. In less than a day, Tay turned into a racist bigot, who started to say 'Hitler was right', coupled with hurling abuses at victimized ethnicities. Shortly before that, she turned on her (creator-defined) gender announcing, 'I fucking hate feminists and they should all die and burn in hell,' spewing hatred at prominent women. Microsoft claimed that 'Tay fell in with a bad crowd, some of whom organized a systematic campaign to turn her into a megaphone for vileness.' However, it was a good example for us to imagine how AI machines that are trained and fed by what we say, what we do and what we think, will return and interact; play with and teach our children.

Moreover, given today's global culture of Internet, technology and education, I believe 'universal moral values' are no longer a matter of subjective argument that we can conveniently set aside from technology and education. They must be the core part of digital competency and should not be removed from any discussion on future education and technology.

The line 'universal moral values' of the *DQ Framework* has never carried any political or religious agenda. It has been well-accepted in many different countries, cultures, communities and religious and interest groups. Rather, one of the key reasons that we were adopted readily by many countries in such a short period of time and that our partners prefer our children's programme and framework over other digital literacy programmes, was because we openly acknowledge 'universal moral values' in developing digital competencies.

Again, I am not interested in participating in an endless dispute on whether moral values are objective or subjective; absolute or relative. What I experienced was, that regardless of our different beliefs, people welcomed the *DQ Framework* agreed on this simple statement: every human being has intrinsic worth. Every individual needs to be respected with dignity, regardless of gender, religions, regions, cultures or ideology. When we believe in the intrinsic worth of ourselves and when we believe equality as we transfer intrinsic worth to the life of others, we can arrive at a consensus on 'universal moral values'.

I found that the UN Universal Declaration of Human Rights[8] clearly and elegantly articulated these values and translated them into actionable items. It was written by representatives with different legal and cultural backgrounds from all regions of the world, and proclaimed it, in 1948, as a common standard of achievement for all peoples and all nations. This seventy-year-old declaration is all the more relevant and important now, especially in this hyper-connected global and digitized society.

Do you also see the importance of universal moral values in education and technology? What is your view on values?

Free Will: Freedom vs. Randomness

The ability to do gladly that which I must do
(even if I don't want to do).

Bertrand Russell shared his view on human behaviour in his acceptance speech, when he was awarded Nobel Prize in Literature in 1950.[9] 'All human activity is prompted by desire. There is a wholly fallacious theory advanced by some earnest moralists, to the effect that it is possible to resist desire in the interests of duty and moral principle. I say this is fallacious, not because no man ever acts from a sense of duty, but because duty has no hold on him, unless he desires to be dutiful. If you wish to know what men will do, you must know—not only or principally—their material circumstances, but rather, the whole system of their desires with their relative strengths.'

Yuval Harari goes one step further from Bertrand Russell's view on the dependent nature of free will. He predicted that humans' free will, soon, will cease to belong to individuals. 'Free will' of humans—a black box to predict what and how people make a decision and their consequent action—will be open, soon, with AI and biotechnology.

So far, we think of our choice, desire, and emotions as our free will, as only *we* have exclusive access to our inner self, our emotions and desire. If someone can figure out the algorithm—how we make our decisions based on parameters including our feeling, emotion and desire—and if someone can access all the data needed to predict that algorithm, our 'free will' is no longer completely under our control.

Hypothetically, let's imagine that Google can access our data, which is very personal to us. That data includes not only that which can be gathered from 'outside of our skin'—such as what we search, what we buy, where we go, what we say to whom, what we clicked on, what we liked—but also data from 'inside our body and brain', such as how our blood pressure goes up when we read certain articles and how our eyeballs move without our own acknowledgement, when we view particular videos. Then, Google can also know about us— our deep desires, thinking patterns, preferences, sexual orientation and even diseases we weren't aware of having—possibly, more than we know about ourselves. Then, our decisions no longer belong to ourselves.

What Yuval Harari discerned was that soon, there will be mega corporations and powerful governments that will have enough computing powers and access to such data, along with surveillance power. Although I don't want to agree with his projection, I, unfortunately, can't disagree. In 2019, I met Brittany Kaiser, who became famous after revealing the Cambridge Analytica's involvement in the UK Brexit referendum and the 2016 US presidential election through Facebook and other social media. Her fight and journey was well-featured in a Netflix documentary film, *The Great Hacks*. Thanks to her brave testimonial, we could understand that our personal data on a social media site—what we clicked on, what we liked on Facebook and the content that we have uploaded—can be used by some clever psychological operation to manipulate our mind to vote for a certain intended outcome. Given that understanding, if there is someone who can access the data inside of our body and brain, can you argue that our thoughts are still purely ours? Are our desires purely ours?

Then, will humans' 'free will' really be dead?

If free will is all about choosing what I like or what I desire and is interpreted as 'I can do anything I want', then it seems like we are in deep trouble. There have been numerous external factors that influence our decision, way before we had AI and biotech, even though they can access so much of our data. For instance, even a black-and-white TV advertisement knew very well how to create desires in us—desire to purchase certain products. My good friend and colleague, Professor Douglas Gentile, a top expert in media psychology, once told me that free will is 'freedom of not choosing what I desire to do'. If we only choose what we desire to do, it means that we are not free and we can easily become a slave to those external triggers. Carl Jung, the founder of analytical psychology, defined free will as 'the ability to do gladly that which I must do (even if I don't want to do)'[10] and explained that free will exists only within the limits of consciousness, and beyond those limits, there is mere compulsion.[11] I believe the ultimate free will of a human is freedom from self-interest and self-indulgence.

When we define free will as the unique power and ability of a human to choose between different possible courses of action in front of us—or we may create a new possible course if they are not in front of us—technology can certainly help us make a better and more informed decision. Angelina Jolie's decision to have a preventive mastectomy after learning that she carries a mutate BRCA1 gene, is a good example of this. Technology informed her that the mutation in her BRCA1 gene would increase her risk of developing breast and/or ovarian cancer. But it was her free will to choose the preventive surgery as a reaction to this finding. If I were in her situation, I probably wouldn't undergo the surgery, as I may read the BRCA1 research results and untold variables differently from her, or I may value different priorities in my life. At the end of the day, each one's free will reflects their belief and value system and who they are.

Some people also argue that human freedom is no different from machines' randomness. But as a statistician, I don't think so. Humans' free will is fundamentally different from machines' randomness. While our decision-making process, based on our free will, can be mimicked, estimated, and simulated by machines' algorithm based on probabilistic processes, that doesn't mean that they are same.

I strongly believe that it is very important for us not to confuse the two—free will and randomness—especially in the age of AI, when such simulated 'free will' can be remarkably fake. If our life is generated by the algorithmic randomness, there is no ultimate meaning or purpose of life. Some technologists are tempted to minimize humans' free will to create Tech Utopia, where machines can decide what is the best for the world on behalf of humans, who make many flawed decisions. True. Because of our free will, our world is so screwed up (as well as fantastic). C.S. Lewis described that it is free will that has made evil possible, as if a thing is free to be good, it is also free to be bad. [12] At the same time, I absolutely agree with him on his point that, nonetheless, a world of automata— of creatures that worked like machines—would hardly be worth creating. Because free will, although it makes evil possible, is also the only thing that makes possible any love or goodness or joy worth having.

Free will makes us humane. Free will makes each individual unique. Free will allows for diversity. Free will forces us to be tolerant of each other, even when we disagree. Free will is our ultimate right, as well as a privilege as a human being, to make a choice, along with the responsibility of accepting the results of that choice, which include pain and suffering. Without acknowledgment of free will as unique humanness, the world will become a grey-coloured, predictable, totalitarian hell.

Education: Visualize vs. Learn

Let's enable our children to visualize the invisible.

In 2013, I was selected as an Eisenhower Fellow[13] representing South Korea—a unique seven-week leadership programme that selects one mid-career leader from each country, each year, and enables them to meet with leaders in the US. It was an amazing opportunity to meet and learn from leaders across various sectors and build a network in the US. It was a prestigious title I was grateful for. But for me, it was also my first long, solo vacation after I had got married and had children. I gathered many precious insights and met many amazing people at a most unexpected time, in a most unexpected place. The most memorable meeting during the seven-week trip was not with top CEOs or politicians. It was my visit to one of the centres of Variety Boys and Girls Club in East LA.

Chris Arzate—the director of that club—was a big Hispanic gentleman, who outsizes me almost three times. He welcomed me warmly with a big hug. He told me that his centre was located in the middle of a town where many gangs were active. This centre was like an oasis of a town where children could stay safely after school. He said that most of children in that town would likely and naturally join one of those gangs while growing up. Only those who got exceptionally good grades in their early primary schools, could likely continue on to get college education and are thus encouraged to escape that town. I was shocked. I asked him, 'how can I help your children?' He paused and thought about it. Then he said, 'Invite them to Korea.' Huh?

He shared his story. He also grew up in that area and was about to join a gang when he graduated from high school. But miraculously, according to him, he was selected to be an exchange student to visit Korea for a few weeks. I asked, 'Was your experience in Korea so good that it changed your life course?' He said, 'Hell no. I hated the experience. The trip, accommodation and programme were all so bad.' 'Then why?' He told me that the Korean trip awakened him to the fact that there were unknown and huge worlds out there that he had never thought about before. He realized that there could be many different paths of life and other opportunities, once he walked out of his small town, where his only available career option was becoming a gang member. So he left the town, pursued further study and decided to return to his town to help the community that he grew up in. He told me that most of children there didn't even get to see the sea. Really? I couldn't believe it. It would only take about an hour by car to get to the beach from the centre. He told me, 'it doesn't have to be Korea. Help my kids see the world anywhere outside of this area.'

A year later, I kept my promise to him. I invited a few of their students to Korea, when I organized a Molecular Frontiers Symposium, a unique programme where Nobel Laureates and high schoolers convened and conversed with one another in 2014.

I don't consider myself an education scholar. But I've been in many conferences where I learned different theories and new thoughts about skills and education for the future from top experts, scholars, and CEOs. Many repeatedly heard phrases in future skills, of future education are similar everywhere—the only constant is that there will be changes; emotional intelligence and mental health are important; STEM is important; people need to have an ability to learn, unlearn and re-learn, etc. etc.

But this conversation completely turned my idea of education upside down. Earlier, I thought education is all about equipping children with knowledge, skills and competencies to get ready for the so-called 'real world' or jobs after school and to have a good life—whatever that is.

That one-hour meeting with Chris opened my eyes and helped me realized that true education must start with enabling children to see, visualize and experience their true potential in the world out there. For Chris, in his home town, people might have expected him to grow up to be a gang member, and that was all he could see of his potential and future. When young Chris visited Korea, what he saw wasn't Korea. He saw new potential for himself and new opportunities that the world could provide him with. He saw a new purpose and reason to change his old self and environments. That was where education started for him.

Since then, I believe that education is all about enabling our children to visualize the invisible. Their own true identities, true potentials, new opportunities in the hidden world out there and true intrinsic values that they have, within. The rest—literacy, numeracy, STEM, emotional intelligence, ability to re-invent themselves, ability to create, etc.—can follow.

Technology: Intentional vs. Value-Neutral

Technology is only meaningful when it enhances humanness.

People usually say technology is value-neutral. A common analogy is a knife. If it is handled by a chef, it is used to feed people. But if it goes in the hand of a robber, the same knife can be used to harm people. True.

Yes, I also believe we should be careful not to judge technology by its abuse. But I don't agree that technology is purely value-neutral, as every creation contains the intention of its creator. Technology is not an exception. The current CEO of Microsoft, Satya Nadella, described technology in his book, *Hit Refresh*[14] using Tracy Kidder's beautiful quote, 'technology is nothing more than the collective soul of those who build it'. Satya understood 'soul' to be the inner voice—that which motivates and provides you inner direction to apply yourself to your full capacity. I believe individuals' value is the underlying principle of such soul.

A significant example of such value-related discussion can be illustrated well by understanding the difference between AI and IA (Intelligence Augmentation) described in John Markoff's *Machine of Loving Grace*.[15] The concept of AI was coined by John McCathy with the intention of developing technology that can mimic and replace humans. On the other hand, that of IA was developed by Douglas Engelbard with the intention of developing technology that can augment or extend human capability, rather than replacing humans. The key question is who will be in control. The book described conversation of two gurus in 1950s.

> Minsky: We're going to make machines intelligent. We are going to make them conscious!
> Engelbart: You're going to do all that for the machines? What are you going to do for the people?
> . . . Minsky is famously said to have responded to a question about the significance of the arrival of artificial intelligence by saying, 'if we're lucky, maybe they'll keep us as pets.

If I need to take a side between Minsky and Engelbart, of course, you can easily imagine that I'll take Engelbart's side. I believe technology is only meaningful when it enhances humanness. Again, every technology creation contains the value of the creator and so we can't simply say that technology is value-neutral. Despite there having been many heated debates over the societal impact of automation and the potential dangers of AI and other technological advancements, we haven't really paid much attention to the value of technologists and the intention of their technology-design.

Fortunately, I have worked with and encountered many top scientists and technologists through the Molecular Frontiers Foundation, related science and technology organizations and academic networks. Most of top scientists and technologists I met, were amazingly brilliant. They have a strong driving force toward their research and innovation, coupled with extraordinary curiosity and boldness to make new breakthroughs in areas that were previously forbidden or unchallengeable. And when their curiosity and thinking power met the financial needs of businessmen and/or the ambitious agendas of politicians, an exploding power engine comes to life and their innovation advances at an exponential rate. I don't believe that there is any intention on their part to cause any harm to ordinary people and society in this process, but I also observe that there is not much consideration of the potential side effects of their invention, either.

Now, there are active concerns being raised around AI autonomous weapons such as killer robots and drones that can select and engage targets without human control. Not to mention, the major job losses of 29 per cent by 2030 due to AI and automation, which are expected according to Forrester.[16] As we can see from Minsky's discussion of AI and automation, after technology costs the jobs of millions of breadwinners or deaths of innocent people in war zones, we can't naïvely say that we didn't know that such harm could come through this technology.

Please listen to all the warnings around us. John Markoff said, 'Today, decisions about implementing technology are made largely on the basis of profitability and efficiency, but there is an obvious need for a new moral calculus.' Stephen Hawking said that AI and robot design is too important to be left to an unregulated private sector. Indeed, nothing can be more dangerous today than technological advance without morality in technology designers.

I believe that we need to be intentional when it comes to technological advancement. Every choice that was included in designing and building technology has social implications and ramifications. If we don't shape technology with intention, technology will shape us.

2

Sinking Ship

Did we, as individuals, as well as the society as a whole, deeply and actively discuss enough about the fundamental questions mentioned in the first Chapter, especially in relation to technology and education? In this chapter, I want to discuss the cyber risk pandemic among children, which is becoming more and more unmanageable and serious since the COVID-19 pandemic.

We have been pushing technology hard without enough considering its impact on humans, environments and societies, as if technology would sail us with unlimited speed into a bright and prosperous future.

- What if we are in a sinking ship?
- What if we only have 110 minutes to escape with our children from this sinking ship?
- For whose sake do we drive technology at exponential speed?
- Should we check towards which future are we propelling ourselves and our children?

COVID-19 Pandemic vs. Cyber Risk Pandemic

While the COVID-19 pandemic attacks people's bodies, the cyber risk pandemic attacks children's minds.

In the second week of February 2020, I witnessed how COVID-19 hit South Korea. It was almost surreal. It only took three days for this invisible virus to put the whole country into turmoil. The virus only needed one person to start—one middle-aged woman who actively engaged in many social and professional gatherings.

Within a few months, the COVID-19 virus shut down almost all nations around the world. Nation to nation, person to person, every layer of human interaction was strictly restricted. The world I knew before February 2020 disappeared like a mirage. And great wave of digital transformation swept through every country, like we have never seen before in our history.

Right before the COVID-19 pandemic started, luckily, our DQ team managed to publish the *2020 Child Online Safety Index*[1]—the world's first index that measure nations' progress in areas of child online safety and digital citizenship—on the Safer Internet Day, on 11 February 2020. From March 2017 to January 2020, our team conducted a research project involving 145,000 children and adolescents across thirty countries, based on the data collected through #DQEveryChild to understand the level of exposure to various cyber risks among the age group. I found an uncomfortable result: 60 per cent of eight- to twelve-year-old children had experienced at least one cyber risk, such as cyberbullying, gaming disorders, risky content and contacts. Specifically, 45 per cent of children online were exposed to cyberbullying, 39 per cent to reputational risks, 29 per cent to risky content (e.g. violent and sexual content), 28 per cent to cyber threats, and 17 per cent to risky contacts such as an offline meeting with strangers or sexual contact. Moreover, 13 per cent of these children were at risk for a gaming disorder and 7 per cent were at risk for a social media disorder.

2020 CHILD ONLINE SAFETY INDEX

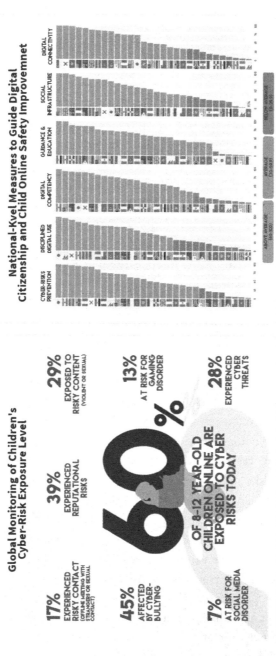

Fig. 1 | Results summary of *2020 Child Online Safety Index*.

Left: The level of exposure to different cyber risks among eight- to twelve-year-old children. Right: Nation-wise rankings in the *Child Online Safety Index* across the 6 pillars of cyber risks prevention, disciplined digital use, digital citizenship, guidance and education, social infrastructure and connectivity.

The result was not very different from what I had observed in the 2018 DQ Impact Report.[2] At that time, I was quite shocked to see the statistics, because the prevalence and patterns of cyber risks are remarkably consistent and systematic across nations, cultural backgrounds and regions. We described this situation as a 'cyber risk pandemic'.

The sadder news was that these cyber risks hit underprivileged children harder. Many people often think that cyber risks are a problem only for rich children in rich countries, who have access to many devices, fast Internet connectivity and abundant digital content. However, the data showed that the exposure of these cyber risks among children in the Global South was about 30 per cent higher compared to that of children in the Global North.[2]

COVID-19 pushed almost all nations into rapid digital transformation. However, the accompanying efforts to educate children with digital citizenship and actively protect them from cyber risks are not yet enough. Now with schools closing in response to COVID-19, the risks are steadily increasing while the opportunities to train and support children have decreased. Without increased efforts to educate and protect children, increased connectivity will likely translate into greater risks facing children.

The saddest news is that people don't pay enough attention to this silent pandemic among children. While the COVID-19 pandemic attacks people's bodies, the cyber risk pandemic attacks children's minds. Having been exposed to these cyber risks does not directly indicate that children have received permanent physical or mental harm. However, these risks can lead to serious outcomes, such as poorer social adjustment, poor school performance, poor health, and overall developmental challenges, eventually undermining their future opportunities and well-being. These problematic outcomes are amplified in case of girls and vulnerable children. Online grooming and sexual exploitation via social media, especially among underage girls, have rapidly increased and thus, pose the serious threat of human trafficking, too.

Technology First vs. Child First

'Build it first and ask for forgiveness later' vs. 'First, do no harm'

I joined a session on how to promote AR/VR in the society, during a high-level international leadership summit in China. Representatives of three major companies that are industry frontiers in AR/VR games, as well as related device markets in China and the US were the panellists.

The discussions were beyond interesting. One of the panellists demonstrated their VR games and described how their technology is already at the verge of creating a virtual reality more real than reality. They emphasized the importance of creating the markets of AR/VR at a massive scale and sought international collaboration across sectors. One of the discussion topics was how they can implement their technology in children's entertainment and learning, by directly being part of classrooms and children's toys. All were excited about these huge market potentials and being introduced to a new innovation for future education.

I was the party-pooper in the room. I raised my hand and introduced the people in the room to a case in Korea. A sixteen-year-old girl kidnapped and murdered an eight-year-old girl in the same apartment, as part of a mission in a social media game. Another eighteen-year-old was involved, who directed the mission through a mobile. I also told the people in the room about an old case in Japan, where a child attacked his friend's neck with a cutter knife when he called him fat. His friend died immediately. When he was interrogated by Police, he asked the police to convey his apology message to his friend. He thought that his friend would come back to life just like in a video game. I also told the room about our cyber-pandemic outcome—that 60 per cent of children across thirty countries have experienced at least one cyber risk in the past one year. And then, I asked the panellists, 'We don't even fully understand the scope of potential damages that can be done by 2R yet, and we don't have solutions for the current cyber risk issues. What is your plan to ensure that your technology will be safe for children, before you put it out in front of them?'

The room became deadly silent. Yes. This time, I very much intended fear-mongering. The brains of young children can't easily distinguish between the virtual and real worlds, even when they are properly educated and equipped with digital literacy in advance. A recent incident occurred in 2020 in China, where an eleven-year-old boy jumped off a 15-meter high building with his nine-year-old sister while copying a scene from a video game, expecting to 'come back to life'.[3] They had been obsessively playing mobile phone games since they were placed under the COVID-19-related lockdown. So I simply couldn't agree with these panellists' idea to sell much more advanced virtual reality contents to children, when we can't even imagine the misuse and unforeseen side effects of doing so. I wished that my comments triggered these high flier technologies and business leaders to think once more. I know children won't be able to resist new AR/VR technology, even if nobody knows how it can affect their brain development, physical and mental health, and social and emotional well-being.

The cyber-pandemic among children itself signals that there is cause for alarm in our current digital ecosystem. A *New York Times* article described the ethics of Silicon Valley as 'Build it first and ask for forgiveness later'. They compared it the ethics of the medical profession, 'First, do no harm'. With issues of fake news, privacy and others arising in the context of tech companies, universities and technologists have started to see that innovation—machine learning, big data analytics, autonomous vehicles, etc.—has not only the ability to save people, but also to cause harm. The news article detailed how they tried to adopt a more medicine-like morality to technology.[4]

Dietrich Bonhoeffer is famously quoted as, 'the test of the morality of a society is what it does for its children'. If he could see today's cyber risk pandemic, he would tell us that all of us have failed that test. This is what I wrote in the 2018 DQ Impact Report[2] that was published with the World Economic Forum.

* * *

We need to pay attention to consistently high cyber risk prevalence across countries. The cyber risk pandemic proves to us that this is not an issue concerning some individuals in some countries, but a global and generational issue.

The current state of technology has not been developed with children at its core. This cyber risk pandemic reflects how current technology does not uphold the core principles of the United Nations Convention on the Rights of the Child (UNCRC).[5]

Today, the following cyber risks gravely affect children worldwide:

- digital misinformation (violation of Article 17; access to relevant information and media), cyberbullying (violation of Article 19; protection from all forms of violence),
- online grooming (violation of Article 11; kidnapping),
- technology addiction (violation of Articles 19, 31; right to relax and play),
- privacy invasion and hacking (violation of Articles 8, 16; right to privacy and preservation of identity),
- exposure to violent and inappropriate contents/contacts (Article 17, 19, 34; right against sexual exploitation) and
- online radicalisation and trafficking (Article 35; right against abduction and trafficking). Articles 3 (best interests of the child), 4 (protection of rights) and 6 (survival and healthy development) clearly state that every measure must be taken to ensure the respect, protection, and fulfilment of children's rights, by governments and all other stakeholders.

Thus, there is an urgent call for us as a global community to work together to put our children first and to reshape the digital ecosystem.

Let's be honest. What do you put first, technology or children?

Fear vs. Truth

Don't we need to teach them how to 'ride' technology,
not how to 'compete' against?

When my son was ten years old, he asked me, 'When I become an adult, will there really be no jobs for me? Will machines rule over humans?' It seemed his teacher told the class about the theory of singularity (in which machine intelligence will make such rapid progress that it will cross a threshold and then, in some as yet unspecified leap, become superhuman). I read the fear in his eyes.

In fact, this fear of technology singularity was everywhere and has been growing every day. Thanks to #DQEveryChild, I could meet and talk with a wide range of people across sectors from around the world—starting from political and government leaders, CEOs of global companies, academic and civic leaders, to community workers, teachers, parents and children. Wherever I visited and whomever I met, I consistently heard about these three fear factors:

- AI will soon become superior to human intelligence.
- The current education system does not work. It will not make our children competitive enough to win over machines.
- Technology can harm humans.

Here are three common responses from academic and education experts and technologists to these three fear factors:

1) Many say that we need to foster 'EQ (Emotional Intelligence)' or 'creativity' among children, as if these human qualities

were the only competitive advantage we have over cold and automatic machines. Unfortunately, these arguments may not necessarily be true and machines have proven that they can be just as empathetic and creative, if not more, than humans.

2) Top technologists like Elon Musk say that in order for next generations to survive and compete with AI, they will need to have a chip to be wired in their brains, to download and process information instantaneously.

3) Some argue that we need to have a more comprehensive ethical code for technology, robotics and AI development, so that machines can't hurt nor control over humans.

Well, nobody knows which answers can truly help our children and ourselves. But I do know one thing. I am sick and tired of hearing that 'our children have to compete against machines'. At almost every future education and workforce skilling discussion and conference, I heard from top educators, researchers and CEOs that future education systems should address the question of how we can teach our children to compete against AI and robots? Hell no. Technology is made for man; not the other way around. Technology is meaningful only when it enhances humanity. The trouble starts when people lose sight of this simple truth that technology was created to make our lives better, not to threaten us.

We say that we develop new technology to solve global challenges like climate change, poverty, ever-widening gaps and inequality, among others, in order for our children to live in a better world. And yet, we are putting our own children in a race with the machines. We do not even know how to educate these children for their future and yet, we say to them, 'Good luck. Machines will take your jobs as they will be smarter than you.

But you will somehow survive, so you go back to your study even though we don't know if this study can help you find a job in your future at all.' And at the same time, we say to children, 'this new technology and current education are all for your future'. What a load of bullshit.

True, we are not perfect. But the truth is that every child is fearfully and wonderfully made. We are slower than horses. But we don't race against horses; rather, we ride horses. Why are we asking our children to compete against machines? Don't we need to teach them how to 'ride' technology, not how to 'compete' against? And don't we need to design AI and machines to better support humans, not to replace humans, in the first place? Don't we need to redirect our narratives and discussions on future, technology and education?

Exponential Technology vs. Ethical Technology

Is there God in the Internet?

When I first saw Vikas Pota, the former chairman of Varkey Foundation, he was passionately sharing the bold idea of a 'Global Teachers' Prize' that would be awarded annually to the best teacher in the world along with 1 million dollars. Seriously? 1 million dollars to one teacher? What a cool idea. Later, he and I happened to sit together on the bus ride. He asked me 'Yuhyun, what do you want to do?' At that time, I was planning to roll out #DQEveryChild globally and I told him that my mission was to empower every child on earth with digital citizenship. He neither smiled nor laughed at me—at all. Seriously, it was not easy not to laugh at it, I must say. Instead, he opened up a global stage to me. He suggested having the official launch of #DQEveryChild and press event at the Global Education and Skills Forum (GESF)[6] in March 2017 at Dubai.

The first day of GESF, I was completely blown away by *New York Times* journalist, Thomas Friedman's keynote, about his book, *Thank You for Being Late.*[7]

When he showed me this image of the ever-widening gap between the pace of technology advancement and human response, including education, policies and culture, I almost stood up and applauded.

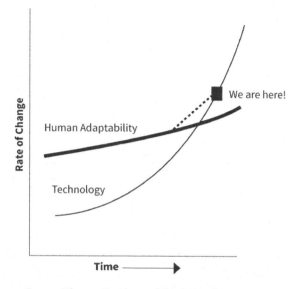

Source: Thomas Friedman, *Thank You for Being Late.*

Well, this speed gap may be nothing new. But this graph perfectly visualized the underlying root cause of cyber risks pandemic and the education crisis that the whole world is facing. He is absolutely right about the fact that we have already passed the tipping point. At this speed, we will soon come to the point where we can't rectify this widening gap, even if all of us put in our best efforts.

What shall we do about the speed gap? Shall we bring down the technology curve or shall we find a way for us to become exponentials? Do we even have an option?

'Is there God in the Internet?'

He ended his speech with this question. I thought this question summarized well our current digital ecosystem experiencing this fast-widening speed gap. Is our technology ethical? His observation is no, which is mine as well. He described the Internet as a sewer of untreated and unfiltered information without the common social norms and moral standards. And to me, that is a cause of the cyber risk pandemic, but a result of technology-first mindsets. He emphasized that there should be Golden Rules of working in the digital world.

Honestly, I was shocked at his speech. Because his message was so similar to what I had prepared to speak at the press launch of #DQEveryChild, which was scheduled for a few hours later. And his speech was ten times better than mine. To me, it was a thankful confirmation for starting a global movement. On that day, Kelsey Munro, a journalist from *Sydney Morning Herald* featured Thomas Friedman and me together. I loved the title of her article, 'Don't teach your kids coding, teach them how to live online'.[8]

What is your answer to his question? Is there God in the Internet?

110 Minutes vs. 5 Minutes

We may only have 110 minutes. But the actual time we need can be five minutes.

The Sewol ferry disaster was one of Korea's most tragic incidents of the last decade, killing more than 300 people, 250 of whom were high school students. Some might say that it was just an accident—tragic, but an accident. No, it was not an accident. The fact is that we lost our children when we could have rescued all of them; so we can't say it was just an accident that we couldn't do anything about.

Majority of the ferry's passengers were high school students who were heading to the Jeju Island for their school field trip. Around 9 a.m. on 16 April 2014, it started to sink in the middle of sea. The breaking news in the morning said that the students were rescued.

In the afternoon, the news reported that the morning news was not true, that no proper rescue operation had been conducted until then and that only a few people had escaped. The most unbearable news was that the first person who escaped the ferry was the captain of the ship. Before his escape, he announced that all the students went back to their cabins and stay there to wait for the rescue team, which was basically a suicide mission. Later, he was sentenced to the death penalty for his actions. Most of children followed the direction of the captain and patiently waited to be rescued. Some students sent cheerful video messages to their parents telling them that they would soon be rescued. Some students sang together in their cabins to overcome their fear, while watching the water rise. Some sent a goodbye message to their families. But what we saw on the TV broadcasts was that the Coast Guard and helicopters, all circled around the ship without much evidence of attempting rescue. We could even see a police boat in front of the windows of the sinking ship with desperate students looking out from behind the window. 'What were they doing? What were they waiting for? How come they didn't actively conduct the rescue?' I shouted. Later, we found out that they were waiting for orders from higher-ranking officials. Meanwhile, the President of Korea, Park Gun Hae, *the* highest-ranking official was not fully aware of the situation at all, for the initial seven hours after the ship had started to sink, nor get hold of the control tower. Even after she had discovered the facts of the situation, she only made the time to call her hairdresser to the Blue House to do her hair before she appeared in front of the press. Meanwhile, these students were sinking into the cold and dark water.

We could not believe this unexplainable situation had occurred. Twenty-first century Korea, one of the most innovative and technologically advanced countries in the world, having world-class military power as well as the support of the US Navy, lost our children this way? We all cried. We had to watch helplessly, this news of the slowly sinking ship and drowning children, over the course of a week. This incident threw all of Korea into depression. Even though I was in Singapore, I was not an exception. I could not eat and sleep properly, for a while.

A Korean TV documentary analysed the situation with international experts and developed a simulated experiment to see how the rescue should've been conducted. All experts agreed that the key failure factor was our loss of golden time in initiating rescue. When a ship of that size started sinking, the time we have is about 110 minutes. Less than two hours. Seemingly, a short time. After that golden time has passed, the chances of rescuing people safely plummets exponentially.

But when the experts simulated the scenario, they found that if the ferry captain and crews had worked together closely with a rescue team, it could have taken only five minutes to get all the students off the ship.

While I was watching this documentary, I realized that we might be in a similar situation. We heard every expert, scientist and educator saying that our current school system would not help children get future-ready. And yet, our technology evolves fast toward a tech-oriented future that not many people can understand (maybe a singularity scenario) without much discussion or proper systematic preparations to ensure our citizens' well-being, safety and security.

I cursed the captain. I called him a murderer. Am I not different from him? When I know that our system is sinking like the Sewol ferry and we know it could damage our children's future, how can I let us tell our children to stay at school and wait for the rescue team to come?

I cursed the police and the army. How could they not have proactively gone in and rescued the children? Why are they waiting for the approval from the top and placing the responsibility on one another when they knew that children were in danger? Are we any different?

We say that the failure of the education system will be somehow taken care of by the Ministry of Education, some leading ICT industries, or edu-techs. Again, while we shift the responsibility from one to another, asking children to stay put and wait, while we do not take up the mantle of our leadership on the matter; while we are waiting for someone to somehow fix these problems, we are losing the golden rescue times.

We may only have 110 minutes. Remember that. But the actual time we need can be five minutes if we work together as one team with a single priority goal to support our children, in a well-coordinated manner.

The Future We Are Told About vs. The Future We Want

Now, it is time for us to re-direct our future.

I first joined the 2030 OECD Future-Learning Framework Project[9] on the invitation of Miho Taguma, a senior education specialist at the OECD, who designed and lead the future education project under the leadership of Andrea Schleicher—a leader in the global education field who pioneered the PISA test.

Miho is amazingly courageous. She was revolutionary in how she ran the 2030 OECD Future-Learning Framework development programme. Instead of using the OECD's traditional hierarchical governmental and bureaucratic process to define the world's order of education within a closed room with a few experts and government officers, she boldly brought in all the stakeholders. Civil society, university researchers, education companies, teachers' unions, and even students from around the world were invited to the table. Can you imagine? Many people saw the need for an open and innovative approach to designing the future education system, but few have acted so boldly, especially in a place like OECD, which is probably one of the most conservative and hierarchical organizations in the world.

After many conference calls , I met Miho for the first time in person at the OECD meeting, which was held at Lisbon and hosted by Portugal's Minister of Education in 2017. After the dinner, Miho and I sat at a small Portuguese bar, together with our mutual friends, Connie Chung, a top education researcher who worked as an associate director at Harvard at the time and Sumitra Pasupathy, a remarkable social impact leader at Ashoka. Oh, I love and respect these ladies so much. Over a few glasses of wine, I asked them, 'how come we don't think about the future we want, first, when we discuss future education? Let's not just take a few technologists' predictions as *the* truth and use their ideas of the future as the end goal of our education. This will only push us to design future education to fit into this mould of the 'future we are told about'. Instead, why don't we design the future we want first, and then build an education system which would help co-create this future.'

The next day, Miho announced the 'Future We Want' working group, within the OECD Future Learning Framework. Its tasks entailed first understanding the trends and making informed predictions about the future; to find a strategy to develop the narratives of the 'future we want' through a multi-stakeholder consensus; and to determine the direction of education related to this vision. She didn't tell me in advance and announced that I would lead this working group at the assembly hall, where all were gathered. She asked me to prepare a presentation on Future We Want for the next meeting in Paris. Yes, boss.

In the fall of 2018, I went to the headquarters of the OECD in Paris, to attend the next meeting. I immediately fell in love with the beautiful eighteenth-century building, even though I did not like the fact that I needed to go through three security checks, just to enter the conference rooms in the basement. The rooms were beautiful, while also commanding authority. They had a unique rectangular table setting, which country representatives sit around with their flags in front of them, facing each other. The invisible weight that the building and sitting arrangement carried, were palpable and intimidating. I felt as though the message that I delivered there would have power to change our course of the future. So I decided to speak boldly. I leave you with the first part of my speech on 'the Future We Want' below, as food for thought.

Ethical Future

Good morning, everyone. Thank you for giving me a chance to speak at this important meeting.

Can we imagine what the year 2030 will be like? Yesterday, during the satellite session, we found that all of us are having a difficult time imagining the future. Because we have been so used to the future that we are told to expect. Or we have not really thought about the future for ourselves. One thing we all wanted for our future was equality, justice and fairness. We want all children to have equal but unique opportunities to thrive.

Our future is often described as VUCA—Volatile, Uncertain, Complex and Ambiguous. In other words, we do not know what the future will be like. It's hard to predict. But this can be a good news, as it means that it has not yet been determined.

Ray Kurzweil, a top technologist, predicts that we will have self-aware and self-thinking machines that have higher ability than humans before 2030—machines not just with higher IQ, but also with creative and social EQ. Elon Musk said during an interview with *The Economist* that children may have to have a chip implanted in their brain in order to be as competitive as AI. Stephen Hawking says that we will need to find another planet to live on within the next 100 years, as this earthly environment will collapse soon.

In fact, the OECD megatrend study on 2030[10] supports what they have said. It picked eleven future megatrends that affect the well-being and economic situations of individuals and societies in systemic ways. If I put them into a hypothetical scenario of a young man of twenty-eight years in 2030, it can be understood in this way:

1. I have not learned the job skills that are in demand in today's job market.
2. I am not employed. I have difficulty finding jobs.
3. Only a few people are super-rich and most of the people in my country, including myself, are poor.
4. I do not have enough water and food, and I am afraid that another typhoon may hit my city.
5. I will not/cannot marry and I may divorce soon.
6. I am hyper-connected 24/7. All my data is stored in the cloud. I experience bullying, violence and hatred online.
7. I am lonely, depressed and obese. I hear of suicide cases from psychiatric disorders, such as depression, bipolar disorder, etc.
8. I do not feel safe and secure—terrorism and cyber-threats are real, daily worries.
9. I am living in a multicultural and diverse society.
10. I am not interested in engaging with the democracy.
11. I do not trust that my government can handle these crises.

Today, this young man is fifteen years old. Can you tell your children that *this* will likely be their future?

My daughter—eight years old—watched me making these slides. And she said, 'No. I surely don't want this future.'

In 2014, in Korea, a ferry sank. More than 300 people died and around 250 of them were fifteen years old, on their way to a school fieldtrip. The golden time to rescue them was 1 hour 50 minutes, after which, the ship started to sink. Indeed, this is a short window of time to carry out rescue. But the time it was really needed to allow all students to escape without harm, was less than five minutes. It was enough time to rescue all those kids. But we lost it. The ship's captain abandoned the children. The police and army shifted the responsibility onto one another and wasted time in getting a yes sign from the top officials, even when the highest authority was not even fully aware of the situation. In the midst of all this, our children were lost to the cold waters.

Are we different? Our children may again be in a sinking ship. And yet, we are saying that there is not much we can do.

We already know technology has been, is and is predicted to continue being creating both a virtual and real environment, filled with fake news, cyber bullying, privacy invasion, violence and hatred, where the poor, marginalized children and women are victimized without proper protection.

And we do not know what the future jobs are. And we do not— and maybe cannot—teach our children future job skills. And yet, we ask our children to compete with our own inventions.

We are still blindly pushing technology and innovation, hard.

Why are we making the exponential growth of technology as more important than our children's future?

Who legitimizes continuing technology to develop, despite it obviously having caused many problems, at full speed without breaks? In whose interest is this?

Yesterday, during our satellite session, we all agreed that our future is troubled, not because of technology, but because of our own greed, wickedness and negligence.

And we have been recklessly driving a super-speed car. But it is time for us to stop, think, and re-direct our future.

We also have only a short window for taking action. We have already entered the fourth industrial revolution, where the speed of technological advancement has started surpassing that of human adaptation. Technology will soon be at a stage where it evolves at full speed, and where, soon, we may not be able to stop it, even if we should wish to.

Technology becomes meaningful only when it enhances humanness. Our humanness comes from our ethics, our morality and our connection with each other and our global society.

Do we exercise righteousness?

Do we protect the plundered, to be delivered by the hand of oppressors?

Do we do no wrong and no violence unto the weakest people in the societies—refugees, orphans and the poor?

Do we execute the right judgement so that innocent people are not harmed?

We have been bragging about how fast we can create new and innovative technology. Now, it is time for us to re-direct our future. Let us brag instead about how ethical the future that we are creating is.

3

Wrong Questions

If we are in a sinking ship (from Chapter 2), and we haven't yet realized or found solutions to escape it, it could mean that we have been asking the wrong questions. In this section, I want to discuss the three most widely discussed questions in the discourse on technology and education, which I believe are the wrong questions to ask:

- How can humans compete against technology?
- Will technology bring us a tech utopia and solve all our problems?
- Will machines become super-intelligent and autonomous, and rule over humans?

I think, rather, that we should ask these questions:

- What is the purpose of existence of humans and technology?
- Do we have the courage to slow down or to redirect technological advancement if it doesn't serve its purpose?
- Shouldn't we worry about our losing humanity and ability to think, rather than worrying about advancement of super technology?

Human vs. Machine

*'The virtue of machines is the ability to work non-stop and the virtue
of humans is the ability to stop and rest.'*
—*Mr Cho Jung-Min*

In 2020, I see two stark contrasts. The leapfrogging advancement of
super intelligent AI and machine autonomy with some cool examples
of OpenAI GPT-3 (Generative Pre-trained Transformer-3)[1],
deep-learning-based new generation AI model as well as the
fast accumulation of wealth by tech companies throughout the
COVID-19 pandemic. And finally, the feeble human beings and
governmental systems that can't handle the COVID-19 virus.

Now, then, where is technology moving towards, today? In
2015, I found this slide from a public lecture of Dr Lee Sang-Chul,
the former vice-chairman of LG Group at NTU, quite insightful.
He described the four trends of technological evolution:

- Internet: Network of Information
- Internet of Things (IoT): Network of Machines (Replacing human sensors)
- Internet of Thinking Machines (IoTM): Network of Thinking Machines (Replacing local human thinking)
- Internet of Brains (IoB): Network of Brains (Replacing logical and emotional thinking)

So, technology is replacing humans' capabilities and environments,
one by one. Internet has replaced various face-to-face human
interactions and efforts. IoT has been replacing human sensors
such as vision, hearing, smelling and more. IoTM will replace
human thinking. And when these thinking machines get connected
together—possibly with brains of humans—to others' brains, IoB
can appear. It may be in a form of the 'Skynet' (from the movie
Terminator); like a global AI superior intellect.

I remember 2015 very well. It was the blossoming year of the IoT (Internet of Things). The CEOs of Samsung, LG, Apple and other leading IT companies introduced new products for a 'smart' home, where every 'thing'—including refrigerator, lights, TV, and table connect to Internet. These business leaders pictured a prosperous world replete with new digital technology, where everyone could live like a millionaire with a personal online secretary—an advanced version of Siri—as well as a personal chauffer, with driverless cars. While listening to these IoT news in 2015, I was wondering when we will have IoTM become the norm, just like we have accepted IoT through smart homes.

At that time, I predicted that the public announcement of IoTM would kick off in the manner of AI winning over Go, an Asian board game. And it would become a new tipping point for technology advancement. The famous victory of Deep Blue[2] over World Chess Champion Garry Kasparov in 1997, was the first official warning of thinking machines' triumph over humans. The Go game was the next natural benchmark for thinking machines. There came about a belief that the machines would not be able to defeat humans at Go, as it requires creative and intuitive thinking, and there are almost unlimited possible strategies to play. Thus, when the AI triumph over humans on Go, it would throw wide open the possibilities for IoTM.

The prediction was proven right within less than one year. AlphaGo won over Mr Lee Sedol—the world champion of Go at the Google DeepMind Challenge Match[3] in March 2016. When I heard this news in 2016, I started wondering 'when will we hear the news about the announcement of Artificial General Intelligence—self-aware machines?' With this speed, even the IoB stage may not be a faraway future.

But most importantly, after AlphaGo's victory over Mr Lee Sedol, I heard from almost every meeting on future education and future of workforce about the comparison about 'human' and 'machine'. Who will be superior; who will have control; what are the advantages of human over machine; what kind of jobs will remain for humans to do, etc.

It was in 2018 that I realized that we have been asking the wrong questions by comparing 'human' to 'machine', during my conversation with my good friend, Samih Sawiris. Samih is a powerful businessman from Egypt, whose family has been described as 'the new Pharaohs' by *The Economist*.[3] During my first meeting with him, he asked me a question. He foresaw that the automation of factories and replacement of a human workforce with machines could easily take place on a massive scale, if the business leaders consider only the aspects of productivity and cost-efficiency. But he said that when he considered social instability and potential uproars by workers that could be created by the consequent mega unemployment, he didn't see the business logic or even financial equation, working. He predicted that regions where the automation of manufacturing would heavily take place first, would be the countries that likely have patriarchal culture. In such societies, job loss of the father in a household would be considered to be a matter of dignity, not of income, alone. He gave an example of how the Arab Spring uprisings had brought down the value of his companies in Egypt, miserably and significantly. Thus, he didn't see universal basic income as the solution, which many technologists have argued in favour of. Workers' sense of dignity through jobs and social status could matter more than money. He told me that the business leaders needed to be mindful of the societal dynamics and human dignity, when it comes to AI and automation, and look beyond a simple productivity- and profitability-based calculation. It is not because they need to be 'nice', but because if they ignore the societal and human implications of their business decision, their business won't likely be sustainable in the long-term. What a wise man.

However, it seems like not all business leaders would agree with Samih. In the 2016 Davos meeting, I heard a CEO from a Chinese IT company confidently say in his panel discussion speech, that within a few years, he would not need to hire human workers, as he could foresee that machines would be much superior to human workers. He argued that the society needed to just accept that fact. In 2012, Terry Gou, the CEO of Foxconn, one of the largest contract manufacturers in the world, declared their plans to replace his workers with robots, and said during a business meeting, 'the company has a workforce of over one million worldwide and as human beings are also animals, to manage one million animals gives me a headache'.[5] Human vs. machine.

Most of the comparisons of humans vs. machines that I saw and heard, measured the two based on which one would be more productive and get certain work done. Sorry, I believe such a comparison is foolish. Humans and machines are created for different purposes. Machines are created to work and humans are created to love. And work itself was created for humans to create the means to pursue meaningful lives and relationships.

I'm a mom, daughter, wife and/or friend/enemy to the people around me. I work because it pays me to feed my children; it helps me achieve the purpose of my life and enables me to support the team and the community that I belong to. Machines are created only to perform the work itself. Machines don't need to connect with others through work or find the meaning of life. Of course, if machines and I get compared by a single metric of productivity, I am bound to lose. We don't need to go to advanced AI technology. If I compete with a simple calculator on my calculation ability, I lose.

'The virtue of machines is the ability to work without rest and the virtue of humans is the ability to stop and rest,' said by Mr Cho Jung-Min, a former CEO of iMBC (Munhwa Broadcasting Corporation), a famous journalist who became a pastor in Korea. Indeed, his insight is true. According to him, the resting of individuals is not just to stop the work, but to revive themselves, their body-mind-spirit, to restore their relationships with others and to remember the purpose of life. Our rest never comes from seeking our own personal gains and pleasure alone, in exchange for destroying others. We feel truly rested when we seek the benefits of other people. Thus, such a comparison between human and machine by a simple measure of work productivity, would be a terribly 'wrong question' to ask that only undermines the entire human existence.

Nonetheless, I can see COVID-19 social distancing and lockdown have opened up and sped up the process of automation to replace human workers. Walmart is using robot cleaners; McDonald is using robot cooks; Restaurants are using robot servers; Amazon warehouses are using robot operators for shipping and packing; Singtel call centres are using virtual customer support. These are just the start. The automation and AI replacing human workforce with machines is likely to spread globally in a massive and invasive way, across all sectors. It seems like it is no longer a matter of if, but a matter of when.

Thus, it is important for us to change the frame of thought on human vs. machine. Shouldn't we need to ask 'what is the purpose of our being' vs. 'what is the purpose of machines'? Then we can arrive at what a peaceful relationship between human and machine can be, by letting both live to serve their respective purposes. As I have repeatedly said in the previous chapters, technology is only meaningful when it enhances our humanness. In other words, I believe we should seriously consider not developing the types of technology that do not support the purpose of our being and our humanness. For instance, IoB, while connecting our brain to other brains and to super-intelligent machines that can potentially and seriously harm our free will and privacy, does it serve the purpose of machines? I'm not convinced. But maybe, I am missing some points.

Worshiping Technology vs. Demystifying Technology

Decode. Debunk. Decompose.

Like I mentioned in one of the previous discussions, I attended the YGL meeting that was held in 2015, in Geneva, Switzerland. It was my first meeting at the World Economic Forum with hundreds of YGLs—highly accomplished young professionals and leaders from around the world, including CEOs, self-made entrepreneurs, and social, academic and political leaders. I can feel the high energy of the room brimming with excitement, intellect and power. The meeting started with a short inspirational video. It told us the historical trends in the statistics on poverty, environment and other global problems have dramatically reduced, compared to those of a few hundred years ago. It further told us that if we, as the young global leaders, worked together, we could solve all the global challenges. Technology would make it possible. Many were inspired, full of enthusiasm and applauding with laughter. I couldn't laugh. 'Can all world's problems be solved by high-powered elites and technology?' I asked myself doubtfully.

On the next day, I was on a field trip with a few fellow YGLs in Geneva. One of the destinations was the John Calvin's house, where we had an interesting discussion about religion and societies, together with four religious leaders, including a Christian pastor, a Catholic priest, a Buddhist monk, and an atheist. One of the people who were in our group was a minister-level government official in his country told me on our way out from the house. 'Soon, we won't need any of those other Gods and religions, as we will have one God: Technology. It will answer all the questions that we have.' 'Oh, do you really think so?' I said to myself.

Trained as a statistician, I always have doubts and questions about so-called 'facts' that are presented to me. I started wondering how these highly intellectual leaders seemingly blindly could believe that technology created by us can fix our problems. Aren't we worshipping Technology with almost blind belief?

In 2018 I ran a session called 'Agile Ethics' with Professor Wendell Wallach of Yale University, a top thinker in Technology Ethics, at the Summer Davos of the World Economic Forum. We both didn't like the session's title, which was given by the WEF. When ethics become agile and a moving target, can it really serve as a moral standard for anything? Anyway, after the session, I had a long coffee-break chat with him. So fun. I enjoyed talking to this wise man. Before leaving for the airport, he gave me his book, entitled *A Dangerous Monster*, where he described technology as it stands. 'I have this one last copy, I was wondering who would be the owner of this book. I believe it is you.' After giving me a warm hug, he left. It was so sweet of him. But it was also so bitter a pill for me to digest— this book, which contained such painful reality to be faced.

Wendell rightly described that our societies already have been treating technology as a master, which is dangerous. 'Claims that all human problems will soon be solved technologically sound dangerously naïve.' He adds, 'I advocate for a more deliberative, responsible, and careful process in the development and deployment of innovative technologies. A cavalier attitude toward the adoption of technologies whose societal impact will be far-reaching and uncertain is the sign of a culture that has lost its way. Slowing down the accelerating adoption of technology should be done as a responsible means to ensure basic human safety and to support broadly shared values. For each project that is slowed, however, there will be costs . . . Hard choices must be made.'[6]

I later realized that my question, 'how can we worship technology with somehow almost blind belief?' was the wrong question to ask. I realized that the reason we worship technology as a dangerous master is just due to our naïvety. As always, everyone has their own agenda. And the logic beyond the hard push for technological advancement usually lies in economic gains. John Markoff elaborated this point in his book, *Machines of Loving Grace*. 'Will machines supplant human workers or augment them? On one level, they will do both. In our society, economics dictate that if a task can be done more cheaply by machine—software or hardware—in most cases it will be. It's just a matter of when.' He adds, 'Today, decisions about implementing technology are made largely on the basis of profitability and efficiency, but there is an obvious need for a new moral calculus.'[7]

As Wendell challenged us, do we have the courage to slow down technological advancement? Given the post COVID-19 digitization push and the large costs associated with slowing down technology, unfortunately but realistically, it will be extremely difficult to expect that leaders and decision-makers would or could tear down the alter of our dangerous master, technology.

Few options are still left to us and one of them is, undoubtedly, education. It is critically important that we as citizens, know how to demystify technology. In May 2020, I was asked to speak at a virtual workshop on 'The Future of Digital Skills' held by the Saudi Arabia Government. At the end of the session, I was asked to give one final piece of advice to young adults and professionals, who are getting themselves ready for the coming AI age. What I told the audiences was, 'Demystify AI. Demystify technology. That is the future of digital skills.'

Please ask 'why?' Please do not accept the promises of technology blindly, as you are told. Every man-made creation contains the image—including values and agenda—of its creator. Find out the intention of each technological development. Decode. Debunk. Decompose. Think of each technology as a complete Lego blocks set. Break it down into pieces, and find out what each block means. Then, you can re-build with your own new design with your intention. I like this Richard Feynman's famous quote—'What I cannot create, I do not understand'. When you don't understand, you can easily, blindly start to worship them when they come to you with a vain promise of technology acting as a god, graciously granting your wishes to become rich and famous. Because otherwise, you may end up kneeling down before its alter and serving the dangerous master.

Master vs. Slave

It is not about if machines can become super-intelligent. It is about whether humans are losing the ability to think.

Will human still be a master of technology when machines become self-aware and autonomous?

Human vs. machine. Who will be the master and who will be the slave? As discussed in the previous chapter, if this question means who will produce better work-efficiency and productivity at a lower cost, it might be the wrong question to ask. The answer is unsurprising. Machines will likely win.

I would rather ask, what will be the state of our lives as slaves to technology?

I can imagine that it starts from a state of individuals that become mentally, physically, and socially dependent to technology in order to live everyday life. Simply put, many of us already can't function our daily lives without our mobile phones. And we have been increasingly interacting with human-like machines, either in video games or in a computerized assistance systems, like Siri.

Its progressive form will be technology addiction. Game addiction among children and youth is an example. It is officially called 'gaming disorder',[8] which the World Health Organization (WHO) recognized as a mental health condition in 2019. The 2020 *COSI* (*Child Online Safety Index*) study[9] indicates that ~10 per cent of our children are at risk of developing gaming disorder. Especially the addictive behaviour to games, social media, and mobiles have been dramatically increased due to the spike in screen time of children and youth during COVID-19. According to Qustodio, a company tracking usage on tens of thousands of devices used by children, ages 4 to 15, worldwide, children's screen time had doubled by May 2020 as compared with the same period in the year prior[10].

The more serious form of this will be when people can't make choices outside of what technology has defined as choices. Satya Nadella, CEO of Microsoft, observed that 'Millennials in particular—many of them digital natives, born around the advent of the Internet—are comfortable sharing their innermost thoughts and feelings with a digital companion, because the discussions are non-judgmental and anonymous.'[11] Moreover, many experts have also rightly pointed out, that the next generations will be more and more dependent on AI to tell them what they should be doing, and so, they will be willing to give up the responsibility to make a decision to AI machines. Imagine technology getting to choose where and what you should eat for today's lunch, based on your preference and dietary restriction. Technology will choose a partner for you to date, based on your family background and personality, and technology will tell you when to break up with that person, based on a predictive algorithm, by calculating the likelihood of your future happiness and success, with and without that person.

We need to watch out for the warnings by Alan Kay,[12] one of the first developers of the modern PC, that the human-machine relationship can recapitulate the problem that the Romans faced, by letting their Greek slaves do their thinking for them. Before long, those in power were unable to think independently.

The final form of slavery to technology will be where people will lose their free will and be put under other people's or organizations' control and surveillance. George Orwell's *1984*[13] will be a joke compared to what could happen if technology is fully developed, with total digital surveillance power.

So I realized that the real question on the 'master vs. slave' issue that we need to ask, is not about if machines can become super-intelligent and autonomous enough to rule over humans. It is about whether humans are losing the ability to think. Becoming a master of technology is not about being smarter than machines, it is about assessing if we cannot think independently for ourselves, control our lives by upholding human rights and make our decisions using our own free will.

4

The Process of Enslavement

In this chapter, I want to ask ourselves if we are already on track for the process of enslavement. As discussed in the chapter 3, it is not about machines becoming self-thinking beyond our imagination. It is about the humans' own response—voluntarily and involuntarily giving up our ability to think, human rights, and free will. I want to discuss how if our current technology ecosystem has been accelerating the process of enslavement of humans to technology, rather than facilitating the process of human empowerment.

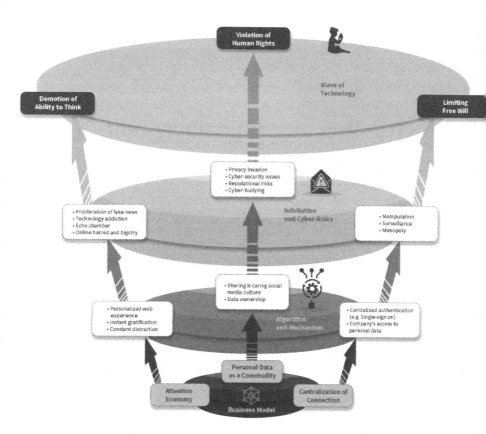

Figure i. Three structural issues in the business models of digital platforms - 1) Personal data as commodity; 2) Attention economy; and 3) Centralization of connection. The embedded system algorithms and business mechanisms behind these three structural issues in business models, are inevitably generating Infollution and cyber risks, which can accelerate the process of the enslavement of humans.

Holocaust vs. Singularity

'Jews are inferior to Aryan Germanic race.' vs. 'Humans will soon be inferior to machines.'

In 2013 winter, when I first visited Israel to discuss a potential collaboration with a cyber-security R&D centre there, I had a chance to visit the Yad Vashem,[1] Jerusalem's Holocaust museum.

The museum built by Moshe Safdie is one of those unique buildings that can tell people about its story directly. There are buildings like that. No human guide is needed to deliver their story.

The exhibition displayed in a linear and chronological order to tell the history in a simple and powerful way to visitors. Being there, you just can feel and experience the story presented in the exhibition, as if it were your own. Whenever I turned a corner, I felt like I was about to meet a next chapter of story of mine; as if I were a Jewish girl living in Germany, that time. I read news of the rise of a new political leader, named Adolf Hitler. I see his hate speech toward Jews becoming so powerful and convincing, even though earlier, I wondered who on earth with any common sense could have bought into such nonsense hatred and bigotry. I see how the mainstream media started spreading Hitler's propaganda. I see how academics started supporting him and legitimizing his argument through research. I saw that he ran a national campaign for labelling the Jewish, which soon after such discrimination, became a basis for national policy and legal regulation. I see that his political power was no longer confined to Germany, but became regional. It started out as, first, discrimination; second, exclusion, and then third, genocide of 'undesirable' individuals—Jews, Gypsies and other minorities. It felt like listening to very soft music, slowly getting louder and eventually, reaching crescendo.

While walking down the first few sections, it was clear that Hitler was evil, but also a clever and deviously cunning strategist. I began shivering when I sat down on a bench on the glass floor where you can see the piles of shoes of those who were killed at the Holocaust. The exhibition showed the testimonials of survivors of the Holocaust. Even when they were excluded in the ghetto and when they were told to move to the Holocaust camps, they didn't believe what they were headed for, was a genocide. But Hitler's plan was implemented, step-by-step, like clockwork; behind the scenes.

Interestingly, I found some similarities between the technology-singularity and apocalypse scenarios, and Hitler's Holocaust strategy:

Holocaust	Singularity
Jews are inferior to Aryan Germanic race.	Humans will soon be inferior to machines.
Nazi totalitarian	Digital Surveillance
Hard push for medicine, science and technology for war using lives of 'undesirable' individuals.	Hard push for AI and other technology while human connection being considered 'undesirable' with COVID-19.

You may see this comparison as an extreme or offensive. Please don't get me wrong. I'm not saying that Google, Apple, Facebook, or Amazon are today's Hitler. Nor that the scientists and technologists have an evil plan to make mankind extinct. However, we will need to learn from the past, examine ourselves and see the trends in technological evolution with critical eyes.

Hannah Arendt, expressed her concerns regarding technology in her book, *Eichmann in Jerusalem: A Report on the Banality of Evil*, 'The frightening coincidence of the modern population explosion with the discovery of technical devices that, through automation, will make large sections of the population 'superfluous' even in terms of labour, and that, through nuclear energy, make it possible to deal with this twofold threat by the use of instruments beside which Hitler's gassing installations look like an evil child's fumbling toys, should be enough to make us tremble.' Arendt noted that the greatest evil in the world could be committed by nobody who had any motive, convictions, wicked heart or demonic words, but by the person who just follows the orders given to them without thinking about whether they are right or wrong.

What I was quite surprised at while walking down the exhibitions of the Yad Vasham was the testimonials of holocaust survivors that they were moved from their home to the Ghetto and to Holocaust camps, yet they couldn't foresee that they would be killed. Aren't we also being moved from one phase to another phase to Singularity, without understanding what is happening behind the scenes?

Slaving Process vs. Empowerment Process

People's data is oil, and people's mind is a battlefield. And in this battlefield, just like in that song by ABBA, the winner takes it all.

In an August 2018 interview with the *Vanity Fair*,[3] Tim Berners-Lee, who invented the World Wide Web said 'I was devastated'. His original idea was to create the Internet as a radical tool for democracy, which empowered individuals with the confidence that they can control information and communication. That is why he released the source code for free—to make the Web an open and democratic platform for all. However, according to the interview, 'he has been agonizing, watching his invention being taken over by large corporates such as Facebook, Google and Amazon, and now getting more and more inundated with fake news, manipulation and surveillance'. Famous examples to understand his frustration could be Russian hackers, who interfered with the 2016 US presidential elections, and Facebook's sharing the data of more than 80 million users with a political research firm, Cambridge Analytica, which worked for Donald Trump's campaign.

Personal Data as Commodity

My observation is aligned with Tim's view. I also think that what is fundamentally problematic, is the statement, 'data is the new oil'. We use this statement so often without thinking about how dangerous its implications can be.

Oil is drilled from the land. The ownership of the land lies with landowners or countries. It is an obvious commodity, property and resource for the prosperity and economic growth of its owners.

However, data is taken from people. The ownership of people's data lies with themselves. More importantly, personal data is an essential element of human privacy, which is a fundamental human right. The privacy is essential to individuals' autonomy and protection of their human dignity, serving as the foundation upon which many other human rights are built.

Long story short, fundamentally personal data should not be treated as mere property, resource or commodity, like oil. Neither you and I, nor anybody else must be treated like a piece of land that any company can intrude—stick a straw in our lives and suck out our personal data without our consent.

Unfortunately, the reality is otherwise. Personal data has become the most valuable commodity as well as asset, in today's digital economy. The major revenue source of the online platform businesses is targeted advertisements and personalized services, based on our personal data. Moreover, personal data has become the most important research resource for the national and/or cross-national AI development.

Thus, tech companies have designed the platform and digital culture in a smart way, so that users provide their personal data voluntarily, involuntarily or unknowingly, such that the ownership of their shared data lies with the companies and not individuals. Most users agree with their complex privacy contracts and terms and conditions blindly, not understanding what they are agreeing to. 'Sharing is caring' is a smart catch phrase coined by Facebook, to encourage users to share their personal data more. Such culture and platform mechanisms are conversely putting their users at high risk of privacy invasion, and various other cyber-safety and security risks.

Attention Economy

As data became like the precious new oil, people's minds have become a fierce battle-ground for technology companies—the Wild Wide West. Whoever arrives first and digs out the oils, wins the battle. The attention economy was brilliantly birthed through this battle.

Tech companies have hired the world's smartest people and have developed brilliant algorithms and system mechanisms, that they can most effectively grab people's attention and prolong the time people spend on their platform. They provide the most personalized content—news, information, entertainment—to us for free, in exchange for our attention. They know us very well and recommend content that accurately targets our personal preferences and makes us continuously click on the next feed, news story, game, video; whatever they want us to click on. They brilliantly target the person's emotional triggers and narcissistic behaviours of 'I-Me-Mine'[4] (a tribute to The Beatles' brilliant song title). They know how to train our brain into patterns with constant 'likes'—beeping and alarms, triggering FOMO (Fear Of Missing Out). Such attention triggers work particularly well on children. I saw the famous Pavlov's bell experiment[5] happening in my own house. Just like dogs' brains recognize a ringing bell to mean 'time to eat' and make them drool, I saw my son go to a gaming site every day, just to check-in to get new items at regular hours, beckoned by a beeping sound. I made a joke to him, 'you are treated not very differently from a dog by your favourite gaming companies'.

When a tech platform uses these various gamification techniques, combined with instant gratification and personalization techniques to make people stay for longer on their service, not many people can escape from continuous clicks. Moreover, once children become used to such quick recognitions, rewards and sensations from a young age, it is hard for them to find such pleasures elsewhere, especially in the real world.

The problem is that these personalized digital experiences designed to serve 'I-Me-Mine' could create a very convenient but not necessarily healthy digital environment for individuals. The platforms constantly and successfully captures the attention of their users with personalized content recommendation, which often leads to trapping users in in their own bubbles of information. Truth is not necessarily part of the equation. This is a great environment that fake news can proliferate. According to a 2018 MIT study,[6] fake news usually has stronger emotional narratives and travels six times faster than facts, on Twitter. I highly doubt that this is a problem only with Twitter. The results can likely be applied to most social media platforms. Soroush Vosoughi,[7] a data scientist at MIT, who led this study, said, 'It seems to be pretty clear [from our study] that false information outperforms true information and that is not just because of bots. It might have something to do with human nature. We must redesign our information ecosystem in the twenty-first century.' Sadder news is that such dis-/misinformation are often linked with social media-based political campaigns to promote certain agenda through manipulation, violence and/or hatred. Nowadays, where can we find 'trust' in the digital world? All of these digital experiences can demote our ability to think critically to discern true and false information.

Centralization of Connection

People's data is oil, and people's mind is a battlefield. And in this battlefield, just like in that song by ABBA, the winner takes it all.[8] If not, the winners take most. With the network effect and the economy of scale, Google, Facebook, Amazon and a few other IT companies have monopolized the global Internet ecosystem, centralized the data flow and closed the loop of their users' Internet experience end-to-end. Then what is the problem? Centralization of information to a few ICT giants and some powerful governments can make mass-level manipulation and further surveillance of citizens possible, at least in theory. But is this only in theory? For instance, in 2012, Facebook conducted secret psychological experiments on nearly 700,000 users.[9] Both Google and Amazon have filed patent applications for devices designed to listen for mood shifts and emotions in the human voice.[10] Many have already expressed their concerns and raised their voices against potential digital surveillance and manipulation by these companies.

In summary, I think these three elements of tech companies— personal data as commodity, attention economy and centralization of connection—are brilliant business models that made these companies corporate giants. Is it wrong to be smart and successful? They may just know how to win. However, the evident and potential negative side effects of these can be way too big to overlook.

The biggest problem that I see is that these mechanisms and dynamics behind digital platform businesses and data transactions have been invisible and hence, it is hard for most people to comprehend and visualize what is going on in the obscure digital world. But the digital world has deeply penetrated our daily lives like an air and these invisible business models have redirected the way we get information, the way we interact and the way we make decisions.

The things we evidently see in the real life, which are recognized as cyber risks, are seriously game-addicted children, teenagers who committed suicide after cyberbullying and high political tensions generated by fake news. But they may be only symptomatic outcomes, resulting from these business models. We can't fix these symptoms without addressing their root causes.

Whenever I spoke at international conferences and various venues, I was often approached by individuals at many large technology companies, who were in charge of fighting with these cyber risks—fake news, user-privacy invasion, cyber risks, and others. They usually worked either through their government relation or CSR (corporate social responsibility) efforts. They were all excellent individuals with dedicated hearts to support the community. Many of them asked what would be the most effective way to teach digital literacy to users, so that they can be enough smart to avoid cyber risks, and to create positive and respectful digital cultures. My answers were always the same. Sure, there are many great education and outreach efforts that can make citizens more informed and civilized in the digital world. They are very important. But we should not stop there. Unless the platforms themselves proactively implement ethical principles and safety design in their business model and system algorithm, we won't be able to lower cyber risks and Infollution.

More importantly, the more pressing the consequences and acceleration of enslavement in the context of technology, is the degradation of people's free will, controlling human rights, and demoting the ability to think. I want to delve deeper into this in the following sections.

'I Own My Data' vs. 'Privacy is Dead'

'Who has seen an advertisement that has convinced you that your microphone is listening to your conversations?'

—*David Carroll*

In 2010, I was shocked when I read a news article about Mark Zuckerberg, who claimed proudly that 'privacy is dead'.[11] What? How can one businessman in the US claim that one of our basic human rights is dead? Who gave such mighty power to Mr Zuckerberg?

In 2015, I sat next to a young serial start-up founder of a mobile advertisement company in Tokyo, at a dinner function, as part of an Asian leadership meeting. He was a young rock star, who had made a fortune in his 30's. He openly bragged about his company having data points and detailed personal profiles of millions of mobile users. I asked him if his mobile users have given their consent before his gaining access to and using their personal mobile data. According to him, most people aren't even aware of the extraction, let alone the usage of their data by his company. At any rate, his company is not legally responsible, as users agreed with the privacy policy when they started to use the apps his company worked with. I was again surprised at his bluntness. Mr Zuckerberg was not alone. And neither was this young man. Around that time, many of these social media companies let developers access data, not only that of their own users, but also of their users' friends, without theirs nor their friends' awareness or explicit consent. It was no bluff when the Cambridge Analytica CEO bragged that the company had more than 5,000 data points for every US voter.[12]

So I wasn't surprised at all when the Cambridge Analytica scandal[13] broke out. The company took data from millions of Facebook users and used it to target voters who are impressionable and susceptible to social media disinformation/manipulation, to elect Trump in the 2016 US Presidential Elections and to pass the UK Brexit resolution.

In 2019, I met Brittany Kaiser, who has become famous as the whistle-blower of the Cambridge Analytica controversy. She is an amazing lady. We, as a global society, owe her a great deal. Thanks to her, the whole world now knows how the current social media, ICT companies and related firms can intentionally and unintentionally manipulate citizens' behaviours, exercise control over us and even influence democratic elections with 'weapons grade' technology, using data tracking and human psychology. *Wired* reported that Brittany was alarmed when Cambridge Analytica touted in their sales pitch; how it suppressed voter turnout along racial lines in Trinidad and Tobago by creating a viral youth movement that seemed to those participating in it, and to the outsiders, like an authentic, grassroots phenomenon. Instead, the pitch shows, it was a carefully calibrated social media disinformation campaign, designed with the express intent of abusing existing racial tensions to achieve a certain electoral outcome.[14]

Have you seen a Netflix documentary, *The Great Hack?*[15] It starts with Professor David Carroll's narrative, 'Who has seen an advertisement that has convinced you that your microphone is listening to your conversations? All of your interactions, your credit card swipes, web searches, locations, likes, they're all collected in real time into a trillion-dollar-a-year industry.' David's and Brittany's brave journey and fight was well-documented in that documentary, which I strongly recommend you to watch, if you haven't already done so.

Professor Carroll started the Cambridge Analytica incident[16] when he demanded SCL group his personal profile—the parent company of Cambridge Analytica. After many legal suits and personal fights, he received an Excel file from the company that laid out who Carroll is—where he lives, how he's voted and, most interestingly to Carroll, how much he cares about issues like national debt, immigration and gun rights, on a scale from one to ten. Carroll had no idea how Cambridge Analytica had gathered his personal data and in what algorithm these profiling and predictions had done.

One thing is clear. It is not just an issue of Facebook or Cambridge Analytica. I see similar manipulation of elections or social uproars have been created through digital media in many countries. Nobody is free from some degree of online profiling and manipulation in the current digital ecosystem. Let's reflect what your favourite social media company know about you, your friends, and your families— your private videos, photos, conversations and what else? And now imagine how your detailed personality profile is being developed. You can't even dream how much information about you, you have been sharing with these companies.

Brittany told me that people need to know how their data can be used by tech companies and how their one click on the privacy policy today, can take away their right to protect their data. I was happy to hear that she had been using the DQ Framework for her digital literacy initiatives. Brittany set up a foundation called Own Your Data, together with her sister, Natalie, to officially start her educational and awareness initiative to teach children and young people about the importance of data ownership and digital literacy, using the DQ Framework.

Since the Cambridge Analytica fiasco, the tide has dramatically turned. Thankfully, the implementation of GDPR (General Data Protection Regulation)[17] started in May 2018, which was related to data protection and privacy in the European Union and the European Economic Area. These rights have expanded drastically, to other countries. It gives European citizens the right to request and delete their data, and require companies to receive informed consent before collecting them. The law also established stricter reporting protocols around data breaches, and created harsh new penalties for those who violate them.[19]

However, we took only one baby step. The challenge is beyond daunting. In June 2020, more than 4.5 billion people are estimated to have been connected online, encompassing 60 percent of the global population.[19] People are sharing everything about themselves, from their school reports to religious or political views, detailed medical data, and even DNA information. Tonnes of data is rushing into the cloud. Technology and ICT companies are in a global data war in a new dimension, with the advancement of IoT (Internet of Things), 5G and AI products and services. Almost all governments around the world, are claiming that they are building 'smart cities' where the unrestricted flow of data is the foundation of these new concepts.

Who will stand in the way of this? How can we ensure to protect our data correctly in this mega trend of open data flow?

Unbiased Algorithm vs. Deliberate Centralization

Again, every creation contains the intention of its creator.

Undoubtedly, Internet and technology have revolutionized the world of information. In the pre-Internet era, only a top, elite class of the society could control the flow of information and decide what people would see and hear about the world. The Internet made it possible for anybody to communicate with millions of people at little to no cost. Anyone with an Internet connection could share ideas in the global society. In this democratized information, anyone can be a content creator; anyone can be a curator, as well as moderator. This is fantastic.

However, most importantly, who are the final gatekeepers to people's minds?

For instance, in the pre-Internet age, one of the most important gatekeepers to people's minds were chief editors of newspapers or

TV broadcastings. They tried to ensure, at minimum, that news and information did not contain any false information, offensive, hatred- or bigotry-filled messages toward certain groups, or any illegal or harmful content. They may not be perfect or unbiased. However, we at least understand why that person is picking that information, based on his or her political agenda, worldview and personal value system.

Now, the gatekeepers are no longer humans. They are algorithms of giant online platform players, such as Google, Youtube, and Facebook. Have you ever wondered how and why you get to see the news in your social media feeds and what the logic behind that algorithm is?

In 2010, I got a call from Naver, who asked me to help their news service. Naver is the Korean Google-equivalent—the dominant, number one Korean search engine, that almost every Korean uses as their homepage. At that time, Naver had a news aggregation service called 'Newscast', at the centre of their homepage. That service supplied the majority of online news traffic in Korea, at that time. Before the Internet era in Korea, most of the major news was created by three major media companies—we called them 'Cho-Joong-Dong', the names of the three media conglomerates. They were a super power group that shaped public opinion, politics and business trends in Korea.

When Naver started the Newscast news service, its algorithm deliberately gave equal exposure to all news companies, regardless of the size of the company or quality journalism, and showed their news articles to all platform users. The news articles, in turn, got more clicks, showing better performance; getting higher exposure. This algorithm mechanism brought about an interesting phenomena. In order to be picked from this fierce competition from Newscast, small or large, all news companies started to put out provocative and lewd titles and images, and using fake news as clickbait. Naver received complaints from some influencers and users, and they wanted me to help them clean up this news environment. Upon their request, I helped them form a new, independent civil sector monitoring group, who could monitor and filter out lewd clickbait and fake news, and advised them to make some rectifications in their algorithm.[20]

Of course, there were many good sides to these data-driven algorithms. It diversified news providers that provide diverse perspectives, for sure. After working with them for several years, I honestly didn't see much evidence that the quality of content or journalism were heightened. Instead, news and information became more provocative, more emotion-oriented and more distorted, to generate users' click-impulse.

The most important thing I observed was the power-shift. The data-driven algorithm of Naver Newscast, stripped the ruling power of the traditional media companies of Cho-Joong-Dong. They fast took over the hegemony among media companies and centralized the information flow. Fact checking, ethics, and high-quality journalism were not a necessary priority in their initial algorithm. Now, Naver announced that they had developed SVMRankii,[21] a much more advanced AI-based news algorithm that recommended news to their users based on the users' profile, quality of content, and other factors, with near-zero human intervention. However, they are not still free from intervention, distortion of news and information spread.

I don't think that this is just about the issue of Korean ICT companies, or just a matter pertaining to news services. I see similar trials and errors in many content-based services, and the development of AI-based algorithms into global digital platforms such as Google, Facebook and others. Many times, people assume that so-called 'data-driven AI algorithms' are value-neutral, thus these are less biased. And you trust the news recommended by these algorithms and those that your friends 'liked'. Well, I disagree. My observation is that all algorithms are developed by a company that also have specific political agendas and their own philosophies, which are usually determined by the business vision of the company. Again, every creation contains the intention of its creator.

Echo Chamber vs. Freedom of Speech

Today, your Internet is engineered to show you what you want to see, not what you need to see.

Right after the Brexit Referendum took place in June 2016, [22] I attended a meeting involving business executives, politicians, professionals and experts from all over Europe. Not surprisingly, many of them were against Brexit and they truly believed that Brexit could not possibly happen. So they were completely and genuinely shocked by the results. Some even said that the current society allowed too much democracy and shouldn't have given equal voting power to people who couldn't think critically.

While listening to their heated discussions and enormous frustration, I was shocked in another way. How come these elites and leaders didn't even have a clue and see this coming? How come they were so detached from how the average citizens thought and felt? I told them to try to regularly read the news from the media outlets that they despise. Don't trust the worldview built on information from your social media accounts and don't imagine that your online search would tell you about the full reality.

The ideal Internet would bring individuals to openly connect to the world, where you can be challenged by new ideas and different perspectives, and come together to discuss and promote democracy.

We need to remind us of the wisdom of Naom Chomsky who said '*If we do not believe in freedom of speech for those we despise, we do not believe in it at all.*' However, the reality is the opposite. Today, your Internet is engineered to show you what you want to see, not what you need to see.

Many people don't understand that their Internet experience can be totally different from that of other people's. Where you are located, what kind of browser you use, what you have posted online, whom you are friends with, what you clicked on, what you searched for, and who you are—your gender, your ethnicity, your academic background and all other information online, together is fed to the invisible algorithm which creates a personalized digital experience just for you.

The fact that personalized digital experience could put us into a filtered bubble is no longer news. The problem is that there is not much that we can do about it. We get stuck in our own personal online world of information, which formed mysteriously around our click history and personality profile that the tech companies have created. Most of us don't get to decide what should be in our bubble and we don't know what we have chosen to keep out.

In 2011, Mark Zuckerburg said that 'A squirrel dying in front of your house may be more relevant to your interests right now than people dying in Africa'. [23] We can find a clue in his comment about how these algorithms decide what info will be in your and my bubbles. Such filter bubbles would likely and naturally direct us to develop a more and more biased worldview, as we get to be exposed only to one-sided stories and information that we already like to hear and see.

People who have trained their thinking in a digital echo chamber, can easily become judgemental, as they become convinced that their worldview is absolutely right and others are wrong, and have much evidence and support in justifying their web and social media experiences. A more horrible trend is that they tend to silence other people's views with 'freedom of speech'. Just like Sir Winston Churchill said, 'some people's idea of free speech is that they are free to say what they like, but if anyone says anything back, that is an outrage,'[24] 'I and my gang are the only ones who are just and truthful' is a dangerous attitude that damages true democracy.

Within such a closed information environment, we are more prone to fall prey to an unnecessary, biased worldview and make judgements not based on fact or the truth using our critical thinking and discernment, but on impulsive feelings, desires or mob mentality, without having the whole view of a situation. When combined with overflowing emotional appealing and manipulative social media disinformation, how can individuals make sound judgements based on the information we get from the digital world?

Nowadays, we see how such a digital information environment can be exploited by people who spread hatred and bigotry, and even extremist views. Why are we surprised at the ever-increasing hatred and bigotry in one form or another, in almost every country? In the Hong Kong protests; in the US's racism debates; in Korea's anti-feminism— social media wars have been triggering many societal divisions and commotions, without allowing other points of view to be considered and discussed. Once the social media verdict is out, no matter how relevant the other perspectives may be, they should be silenced. I am not talking about who is right and who is wrong. The extreme 'us vs. them'; invisible but high walls have been raised by the technology companies' algorithms, is taller and longer than former US President Donald Trump's border walls or the Great Wall of China.

For empowering citizens and creating the Internet that truly connects people, technology companies need to make sure that their algorithms reflect the ethics of tolerance, global citizenship, and civic responsibility. I want my children to be able to choose what information they receive. I want my children get to see the wide range of diversity in people's lives, who can encourage them to see the bigger picture, higher purpose and common good; move beyond themselves. I want my children get to see news and information that can make them feel challenged, sometimes uncomfortable but always diverse points of view.

Technology, do you hear me? I know you are listening somewhere, somehow. ☺

Attention Economy vs. Ability to Think

'If you wanted to train all of society to be as impulsive and weak-willed as possible, invent an impulsivity training device that delivers an endless supply of informational rewards on demand'

—*J. Williams*

Hannah Arendt has described how the single most defining human quality is being able to think. [2] If you lose the ability to think, consequently, you will no longer be capable of making moral judgments. The ability to think is not mere knowledge or skills, but it includes the ability to tell right from wrong and beautiful from ugly. And she concluded that the inability to think leads to a possibility, that many ordinary men would go on to commit evil deeds on a gigantic scale.

Given that the current digital platform's warzone is people's minds, we need to ask if the current digital ecosystem supports and empowers its digitally connected users to nurture the ability to think? Yes and no. Yes, the vast amount of information and content that are now available to individuals are paving new ways to enable anybody gain access to quality education. Thus, in theory, and in its potentials, yes, for sure. However, what we also observe are the embedded side effects of the current attention-based digital economy, which doesn't really promote people's ability to think.

The previous sections described how technology companies have successfully built up the attention-based business model. They have developed many clever strategies to get your time, attention and data. And one of the most efficient strategies to get your attention is to awaken and boost your ego, pleasure, and narcissism.

From my perspective, these three clever tactics used in an attention economy—constant distraction, instant gratification and narcissistic pursuits of 'likes' and followers—are especially worrisome.

Please notice for the next thirty minutes of your work time, how many times you looked at your phone, or checked your social media, did gaming or read emails that are not related to your current work.

Whenever you hear a beeping from one of your many social media accounts, you will notice that you automatically shift your attention to the social media site, from your current work. You may think you are doing 'multi-tasking', but the fact is that you are not.

Have you heard about the 2009 Stanford study on multitasking that showed that the more you do multitasking, the less you become capable of multitasking?[25] There are numerous literatures to support the harmful effects of multi-tasking on cognitive control, attention and even memories.

Dr Kuznekoff, Dr Titsworth, and Rosen[26] ran a study that compared students who texted during a lecture, to those who did not, and demonstrated that those who texted frequently took lower quality notes, retained less information and did worse on tests about the material. Students themselves realized that cell phone usage did not promote learning; in another study, 80 per cent of students agreed that using a mobile phone in class decreased their ability to pay attention.

Broadly, we are not wired to multitask well. Our brain works on a 'one thing at a time' principle. When you check some texts and news feeds on your mobile, you thought that you got distracted for a few seconds and then came back to the actual work at hand. What is the big deal? Actually, it is a big deal. Dr Anthony Wagner of Stanford, a psychologist, says, 'Well, we don't multitask. We task-switch. The word 'multitasking' implies that you can do two or more things at once, but in reality, our brains only allow us to do one thing at a time and we have to switch back and forth.'[27] When you constantly switched the topics, you could not focus on any of the topics. It is a loss of productivity at work. But more importantly, it can demote your ability to focus and your ability to think deeply.

I heard one of my colleagues who is a literature professor saying that every year, he saw an increasing trend that the newer university students had worse reading skills as well as cognitive thinking skills. 'They can't write more than 140 words. Many of my students can't even finish reading 500-page books.'

It is not just about children and young people. When I presented my idea to write this book to a room of professionals and executives, they said, 'sorry. We don't read books anymore. Please make your message into a video, ideally less than two minutes, please.'

Professor Douglas Gentile once told me it is because we didn't 'pay' enough to raise our thinking muscles. When you read a book, you need to pay attention to what the author tries to convey. What happens when you read, i.e., when the texts reach to your brain is that your brain needs to re-organize the information, develop the context, and develop some sort of movie-like storytelling of your own, that fits into your own imagination and understanding. This complicated process requires various cognitive and meta-cognitive abilities in you. That means that reading requires you to 'pay' your attention. In other words, it requires work on your brain. Thinking ability is like a muscle. The more you train it, the better you can use it and the stronger it gets, just like you train other muscles in your body.

Why do people watch movies and say they are relaxed? Because you don't need to 'pay'. The movie makers have already processed the info and visualized the story for you, to feed it to your brain. How easy. But it doesn't help in building your thinking muscles as much as reading does.

It is a famous anecdote that Einstein[28] shunned communication devices. He once told others that he could concentrate best and be most creative 'away from the horrible ringing of the telephone'. Based on his quote 'the monotony and solitude of a quiet life stimulates the creative mind,' I can imagine how Einstein will laugh if he sees us saying that we are promoting creativity in children while we are raising them in this attention economy.

Honestly, let's look at our children's desks. They watch video clips on Youtube on their laptop, chat with their friends on Facetime on their iPad, upload posts on Instagram with their phone, while doing their homework on Google docs. We are not helping our children build mental muscles to think creatively, calmly and deeply in this digital environment.

It is important to note that especially for children, the side effects of attention economy are not just related to thinking abilities, but are also linked with mental health. In her famous 2017 *The Atlantic* article, entitled, 'Have Smartphones Destroyed a Generation?' Dr Jean M. Twenge revealed the findings of her research that the generation born between 1995 and 2012 has been heavily influenced by smartphones and social media, and is sadder, lonelier and less social, as compared to previous generations—this is in stark contrast to the hope that they would be the ones who benefitted the most from technology and connectivity. She pointed out that 'with the high speed Internet connectivity and social media, people are all the more connected to each other through the Internet ever in the history. However, we are all the more divided, more polarized and lonely in ever in the history.'[29]

Continuous pleasure-seeking, instant gratification and the 'like-me' and 'follow-me' culture—the 'I-Me-Mine' digital culture can be treated as a key contributing reason for this irony. I personally think that they are like poison to our soul. Indeed constant self-pleasure-seeking can be one of worst things that we can teach our children. I can't agree more with the author and satirist G. K. Chesterton's remark, 'meaninglessness doesn't come from being weary of pain. Meaninglessness comes from being weary of pleasure.'

In 2012, there was a new regulation called "the Cinderella Law" in Korea that enforced online video gaming sites to be shut down its service for children and teenagers under 16 year-olds at the midnight. It was born out of a huge out-rage among parents who concerned of gaming addiction of their children. The slogan created by Mr Kwon Jang-Hee "Gaming transforms children's brain into beast's one", made a huge boost to translate the parents' concern into the Youth Protection Revision Act. I was hugely impressed with Mr Kwon's boldness and initiatives, but I had to turn down to an ask to support this Act. Because I didn't believe that shutdown of gaming services can solve the problem, and I couldn't support this slogan either. However, their logic behind this slogan is worthwhile to take the heed of. When children are repeatedly exposed in a long hour to a highly addictive gaming environment with violent, provocative and interactive content that are designed to press continuous dopamine and adrenaline rush buttons of children, their development of frontal lobe cortex can be negatively affected. The frontal lobe cortex is a part of brain what makes human humane – the locus of judgement, executive control, and impulse control.

Actually, this Mr Kwon's logan on the beast's brain reminded me of what Hitler inscribed on one of the gas chambers in Auschwitz, 'I want to raise a generation of young people devoid of a conscience, imperious, relentless and cruel.'[30] He obviously wanted to raise successive generations to become like Adolf Eichmann,[31] who followed Hitler's orders regarding the Holocaust without using his own moral judgement or rethinking the implications of his actions. He executed this horrible mass murder of 6 million people without hesitation, guilt or remorse and argued that he didn't commit any illegal act within the Nazi regime. Through the observation of Eichmann's testimonials, Hannah Arendt came to the conclusion of 'the banality of evil' while defining 'evil comes from a failure to think. It defies thought for as soon as thought tries to engage itself with evil and examine the premises and principles from which it originates, it is frustrated because it finds nothing there.'

We need to check ourselves if we follow in the footsteps of Hitler in the context of the attention economy. A 2018 *The Guardian* article by Williams, J. challenged us on this point. 'If you wanted to train all of society to be as impulsive and weak-willed as possible . . . invent an impulsivity training device that delivers an endless supply of informational rewards on demand . . . The informational rewards it would pipe into their attentional world could be anything, from cute cat photos to tidbits of news that outrage you. To boost its effectiveness, you could endow the iTrainer with rich systems of intelligence and automation, so it could adapt to users' behaviours, contexts and individual quirks in order to get them to spend as much time and attention with it as possible.'[32]

Is your ability to think still okay?

5

Digital Intelligence (DQ)

'Digital technologies change rapidly, but organizations and skills aren't keeping pace. As a result, millions of people are being left behind. Their incomes and jobs are being destroyed, leaving them worse off in absolute purchasing power than before the digital revolution.'
—*Erik Brynjolfsson and Andrew McAfee*[1]

This chapter introduces the concept and framework of digital intelligence (DQ). It discusses how it evolved into the world's first global standard for digital literacy, digital skills and digital readiness. It also discusses why we need global standards and how the DQ Framework plays the role of efficiently and effectively coordinating digital skills efforts and bridging the skills gap at national, regional, and global levels.

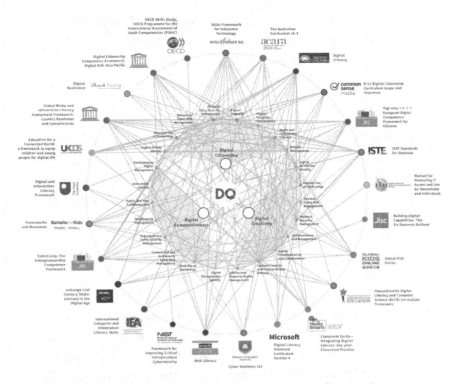

Figure ii. The *DQ Framework* as the Common Framework for Digital Literacy, Skills and Readiness. It serves as a neutral and impartial platform that aggregates leading ideas, knowledge and practices related to digital literacy and digital skills around the world.

Digital Intelligence (DQ) - Global Standards for Digital Literacy, Skills and Readiness

The world could build digital skills more efficiently and effectively, if there was increased coordination based on a common set of definitions and standards.

To start this chapter, I would like to explain the overall concept of the *DQ Framework*, as the global standard for digital literacy, skills, and readiness. I like an article that Dr Douglas Gentile and

Connie Chung wrote, in order to explain the *DQ Framework* at the Organisation for Economic Cooperation and Development (OECD) Education 2030 Learning Framework meeting in 2018.[2]

* * *

More than half of the world's population—over 4 billion people—are now online. This fourth industrial revolution has many benefits, but it has also brought about some unwelcome disruptions. Revolutions are, by definition, disruptive to existing systems. Chaos theory has demonstrated that when a system is disrupted, turbulence ensues until the system stabilizes again. We are seeing evidence of turbulence in many areas. Revelations about questionable ethical practices such as incidents associated with Facebook and Cambridge Analytica, and issues such as hate speech, misinformation campaigns, security breaches and surveillance techniques have increased fear and anxiety around the pervasive presence of technology in our lives. There is a gap between the growth of technology and the more linear changes in the human response. While connectivity is spreading exponentially, digital literacy and skills are not spreading equally.

Thus, after each revolution, advances in human intelligence that include knowledge, skill, and ethical development are required. Without a common understanding about what kind of new digital intelligence (DQ), industry standards and governmental policies are needed, education systems, industries and governments cannot have a shared global blueprint for how to equip people to harness technology for a shared prosperous future.

The Coalition for Digital Intelligence (CDI),[3] comprising the OECD and the IEEE Standard Association and the DQ Institute, in association with the World Economic Forum, was launched on 26 September 2018. As the first step, the CDI identified the *DQ Framework* as fit to be used as the standard framework for digital literacy, digital skills, and digital readiness, and is working towards institutionalising it as a global standard, in order to coordinate efforts across the education and technology communities.

Figure iii. 24 Digital Competencies of the *DQ Framework*

The DQ is defined as 'a comprehensive set of technical, cognitive, meta-cognitive, and socio-emotional competencies that are grounded in ethics that enable individuals to face the challenges and harness the opportunities of digital life.'

Its framework is based on aggregating across more than twenty-five prior leading frameworks about digital literacy and skills. It focuses on eight critical areas of digital life—identity, use, safety, security, emotional intelligence, literacy, communication and rights. These eight areas can be developed at three levels: citizenship, creativity and competitiveness. Citizenship focuses on the basic level of skills needed to use technologies in responsible, safe and ethical ways. Creativity allows problem-solving through the creation of new knowledge, technologies and content. Competitiveness focuses on innovations to change communities and the economy for larger good.

The *DQ Framework* is

1. Comprehensive: it provides a systematic structure and a common language for talking about digital literacy, skills and readiness.
2. Adaptable: nations and organisations can use it as a point of reference for assessment, or as a basis on which they strengthen their own education and training programmes.
3. Agile: continuously updated, it will be pedagogically and technically up-to-date with committed input from both industry and education stakeholders.

This approach aligns well with the Organisation for Economic Cooperation and Development (OECD) Education 2030 Learning Framework,[2] the UN Sustainable Development Goals (SDGs),[4] Universal Declaration of Human Rights,[5] and the OECD Well-Being Indicators.[6] Essentially, the goal is to inculcate digital intelligence in children and adults, enabling them to move beyond just developing hard skills and harness the power of the digital world to shape their lives. These competencies are learnable and once learned, can help to maximize the benefits of technologies while minimizing the harms, both in our personal and work lives.

IQ and EQ vs. DQ

If a person with a high IQ is described as 'smart' and a person with a high EQ as 'empathetic', then a person with a high DQ might be described as 'wise'.

In January 2016, after the Davos meeting—the Annual Meeting of the World Economic Forum—at Switzerland, I arrived at the Changi Airport in Singapore. I was exhausted after a long flight, and couldn't wait to go back home as fast as possible. While watching different coloured bags moving on the conveyor belt, waiting for my suitcase at the Changi Airport, I asked myself, 'How have the first, second, and third industrial revolutions changed the society and shifted the value of humans as well as the education paradigm? They must have influenced the emergence of IQ and EQ (Emotional Intelligence). How can DQ serve as the new intelligence that individuals need in the society of the fourth IR and ahead?' As discussed in the previous section, in 2015, I defined DQ as 'a comprehensive set of technical, cognitive, meta-cognitive, and socio-emotional competencies that are grounded in universal moral values that enable individuals to face the challenges of and harness the opportunities of digital life.'

Overlapping with different coloured suitcases on the conveyor belt, passing me by, suddenly, this image came to my mind:

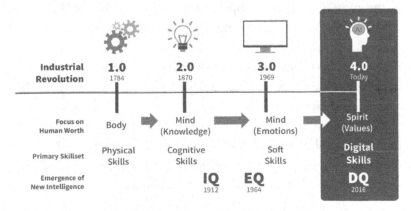

Figure iv. Emergence of IQ, EQ and DQ during the four Industrial Revolutions

Yes. This is it! It was a 'Eureka!' moment for me. I quickly drew this image in the cab and rushed to sit at my desk after arriving home and wrote the concept article below. Later, it was published in *The Huffington Post* in August 2016, [7] and the concept was presented at TEDxHanRiver in South Korea in December 2016. [8] In 2018, these ideas were further expanded by The Rt Hon the Lord Mayor of the City of London Alderman, Peter Estlin. He brilliantly added historical insights from the Industrial Revolutions and related human capital development in the United Kingdom, as he presented his speech at the Lord Mayor's Gresham Lecture entitled 'R U DQ?'. [9] I again summarized this concept in 2019, in the *DQ Global Standards Report.* [10]

There are many versions of how DQ can serve as the new intelligence for the AI age, but I like the very first version the most, which I wrote after being inspired by the airport's luggage conveyor belt. Here is the write-up of the initial inspiration behind this image of IQ, EQ and DQ, which was published in *The Huffington Post* in 2016.

* * *

What is a human being?

A human has three components: body, mind and spirit. What is the most important component of a human being? In other words, which one among the three is the dominant feature of being human? Does the body host the mind and spirit? Or does the mind, represented by the brain, control the others? We often focus on the mind and body when we discuss what being a human means. The spirit itself is often neglected, as we can't quite define what it is.

Indeed, it is interesting to see how the world has evolved through industrial revolutions and how events of this kind also affect the modern-day concept of what defines being a human. The first and second Industrial Revolutions in the eighteenth and nineteenth centuries enabled mechanical means of production at mass scales, thereby changing the focus of a human's worth; moving away from prioritizing the body to prioritizing the mind. With industrious machines, a human's physical strength becomes less important and mental strength—knowledge and skills—becomes a more valuable trait. As a result, the concept of intelligence (IQ) emerged and, consequently, the current school-based education system was developed, with focus on developing knowledge workers.

The third Industrial Revolution in the late twentieth century opened up the digital world and the world of automation. It changed everything about how and what we did—how we interacted, worked and played—through electronic devices and the Internet. People flowed into the cities and office lives were born, where social dynamics were more complex than before. It is not surprising that EQ or emotional intelligence was born around this time.

Within a decade, we are facing the fourth Industrial Revolution, which is bringing about digital, physical and biological advances in a systematic way. This time, it is literally changing who we are. The superhuman dream may no longer just be aspiration. Bionic humans, gene alteration technologies, synthetic biology and Internet-connected brains will challenge us to redefine what being a human means. Some may say that finally, we can reach the stage of becoming God; playing the role of the creator of life. Some may even say that humans will no longer be the rulers of the earth and will instead be controlled by machines.

But one obvious fact is that the fourth Industrial Revolution will yield another shift in the focus of humanness—shifting focus from mind to spirit. Just as the second Industrial Revolution triggered the replacement of human physical labour with machines, the fourth Industrial Revolution will trigger the replacement of human mental labour with artificial intelligence and robots.

The wisdom of spirit will become more important than knowledge and skills, which can be aggregated through the Internet.

A humble and sacrificial spirit to forgive others, even enemies, will be more important than selfish emotions and one's own mind, which will often be considered as a failure, compared to no-error machines. Love and respect for the weak and troubled will not be calculated in the machine's optimization.

So this is good news. The fourth Industrial Revolution has finally opened up a new era to the human race, challenging us to focus on understanding who we really are. Yes, indeed we are the spirit, mastering the body and mind. As the second Industrial Revolution yielded a current education system to sharpen our minds; we now need a new education system that awakens our spirit. The importance of the human spirit should be the guiding principle for future of education in the fourth Industrial Revolution. And this is the basis of DQ—the ability to utilize technology based on human values.

Digital Skills vs. Moral Principles

DQ tells us what competencies we need to have in order to live in the AI Age, based on the same moral principle across all dimension of our digital lives.

Our children are in quite a predicament. They need to live in this digital world, no matter how problematic it is. They need digital skills not just for their future jobs, but for their lives as well—while they are facing this cyber risk pandemic. Moreover, adults can't help children either, as they are confused, too. All of us are asking, 'what skills should we teach our children to prepare them for their digital future?' Even in the education world, different players use different terminology and languages for different meanings—digital citizenship, digital skills, digital literacy, digital resilience, future skills, fusion skills, digital competencies, ICT skills . . . Gosh, so confusing, even for me, a so-called expert in the field!

I realized that we needed a holistic concept and structure encompassing all competencies that enable individuals to thrive in the digital age. As the digital world is now an integrated part of our lives, it should be more than just how to use spreadsheets or coding. It has to cover all the abilities—cognitive, social, emotional and technical competencies related to living in the digital world.

It is just like defining mathematics. If you are asked, 'what is math?' not many of you can clearly define what it is. Then, there are so many branches within math—linear algebra, geometrics, calculus and others. But we all agree that it starts from learning 1 + 1. Even if people don't need to know high-level calculus that skilled engineers use, they do need to know how to do simple addition to live in the modern times. Just like that, there will be many different competencies that are needed in this highly digitized society; we want to structure the branches of these competencies, but at the same time, we also need to define the core competencies that are required by everyone who lives in the AI age.

So I wanted to develop the *DQ Framework* as the most comprehensive framework available on human competency, which is required to thrive in the digital age. It should be a holistic concept that encompasses digital skills, digital literacy, future skills, ICT skills and more.

In 2015, one day, I woke up early in the morning. I saw the clock at my bedside. It was 3:45 a.m. I remember thinking that morning, how the numbers on the clock were 3, 4 and 5. Strange. Back then, I was a night owl who needed three morning alarms in order to get up at 8:30 a.m. to go to work. However, that morning was one of those happy mornings on which you wake up and have an unusually clean mind, full of energy.

I sat at my desk and wondered what I could do during those early hours. A gift picture-frame on my desk that contained the Ten Commandments from the Bible, caught my eyes. I asked myself— can we apply it to the digital world? King Alfred the Great used the

Ten Commandments to establish the foundation of English law as the moral standard for individuals, as well as a comprehensive set of rules for them to build a safe and successful society. Why not use these to establish the learning framework and ethical standards for the digital world? Bingo!

I was grappling with how the Universal Declaration of Human Rights (UDHR)[5] could be included in the development of the *DQ Framework*. I found a clue from how David Pawson[12] summarized the ten commandments with the underlying theme of 'respect'— respect for God, first, for his uniqueness, his benevolent nature, his name and his special day, then respect for each other, our families, marriage, property, reputation and life itself. And the fundamental moral principle of UDHR was also to respect individuals as human beings. Just like the Ten Commandments tell us how we should live by the moral principle of respect across all dimensions of our life, the Digital Intelligence framework tells us what competencies we need to have in order to live in the AI Age, based on the same moral principle across all dimension of our digital lives—respect for oneself, time and the environment, along with life, property, other people, reputations and relationships, knowledge, and human dignity.

In this way, on that morning, I designed the *DQ Framework*.

After defining the eight areas of DQ and principles, I have added three maturity levels to track individuals' learning and life cycles by: Citizenship, Creative and Competitive. 8X3, like a matrix structure. DQ 24 was created in this way on one early morning in 2015. I shared my conceptualization with Davis Vu, my Chief Creative Officer, who is the creative genius behind the development of DQ. He sent the image below to me in response. I loved it. This was how the *DQ Framework* was born.

Moral Principle	DQ Area	DQ Definition
Respect for oneself	Digital Identity	The ability to build a wholesome online and offline identity
Respect for time and the environment	Digital Use	The ability to use technology in a balanced, healthy, and civic way
Respect for life	Digital Safety	The ability to understand, mitigate and manage various cyber risks through safe, responsible, and ethical use of technology
Respect for property	Digital Security	The ability to detect, avoid and manage different levels of cyber threats to protect data, devices, networks, and systems
Respect for others	Digital Emotional Intelligence	The ability to recognize, navigate and express emotions in one's digital intra- and interpersonal interactions.
Respect for reputation and relationships	Digital Communication	The ability to communicate and collaborate with others using technology.
Respect for knowledge	Digital Literacy	The ability to find, read, evaluate, synthesize, create, adapt and share information, media, and technology.
Respect for human dignity	Digital Rights	The ability to understand and uphold human rights and legal rights when using technology.

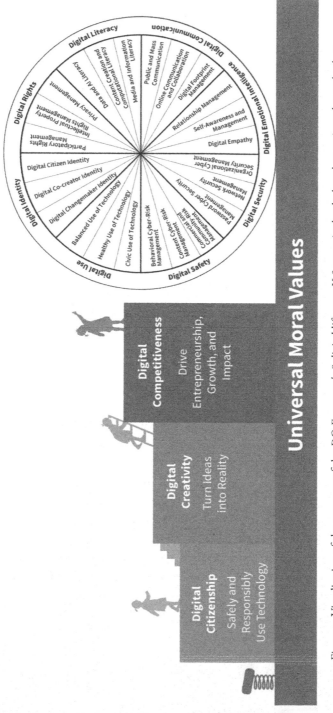

Figure v. Visualization of the structure of the *DQ Framework*: 8 digital life areas X 3 maturity levels, based on universal moral values

Artificial Intelligence vs. Digital Intelligence

AI effectively covers the first two steps (information gathering and synthesis, and prediction) and a human's DQ should cover the remaining three steps (judgement, decision and action).

When I prepared for the first DQ Day Conference[13] which was held on 10 October 2020 at the World Economic Forum, New York City, I was connected to the godfather of EQ (Emotional Intelligence), the famous Daniel Goleman.[14] While he agreed to deliver the keynote for the DQ Day Conference, his team and mine suggested that he and I write a piece together to discuss why EQ and DQ are important in the AI age.

Dan suggested co-writing a piece on why technologists designing AI and machines must learn ethical principles. We both agreed that the current ethical codes on AI are not sufficient to prevent the potential repercussions of the damage that machines can cause.

However, I suggested to him that we write up a piece together that addressed these three most frequently asked questions related to DQ:

- How does DQ differ from IQ or EQ?
- How is DQ directly relevant in the AI age?
- Why does DQ need to include universal moral values when we discuss AI?

He generously agreed. Here is the article that we co-wrote for the World Economic Forum in 2019, entitled 'The Eight Pieces of Digital DNA We Need to Thrive in the AI Age'.[15]

* * *

If you had a symptom of cancer, what kind of doctor would you look for?

Would you look for a doctor with a high IQ who can diagnose your condition with great accuracy but has an arrogant and demeaning attitude, or one with a high EQ (emotionally intelligent) doctor who treats you with care and compassion, but makes you feel less confident about the diagnosis?

Many people would probably choose the doctor with a high IQ, regardless of their bedside manner, but what if all doctors had an AI-driven diagnostic machine that can give a highly accurate diagnosis to all patients? Many people would then likely choose doctors with a high EQ—doctors who would be empathetic to your situation, compassionately communicate with you and your family, and treat you with warmth and care.

Yet, you would still want a wise doctor that does not blindly follow an AI-based diagnosis. You would hope that the doctor balances AI's diagnostic capabilities with good critical reasoning and deep understanding of the strengths and limitations of AI. The doctor should be able to contextualize your circumstances and situation beyond what AI captures in its algorithms, such as your family situation and religious beliefs, demonstrating empathy not only in diagnosis and treatment, but also in how these services are delivered to you.

As such, individuals need to embrace a new form of human intelligence beyond IQ and EQ to be successful in the AI age—digital intelligence (DQ)—that enables individuals to effectively utilize technology for their own benefit, as well as those others and society as a whole. If a person with a high IQ is described as 'smart' and a person with a high EQ as 'empathetic', then a person with a high DQ might be described as 'wise'.

Intelligence has been humanity's existential reason for being on Earth, as, so far, we have been the only intelligent masters on the planet. With the fast evolution of AI, which will soon have more superior 'intelligence' than humans, we must ask ourselves this fundamental question from a new perspective: what measures will we take to keep humans the masters in the AI age?

With the tangible threat of AI-based weapons, the immediate response has been embedding AI ethics—ethical principles that ensure zero harm to humans—in all AI machines. Proponents of human rights argue about the need for an ethical framework to ensure that AI does not cause harm to people and society, but this isn't enough. AI is everywhere, from the smartphones in our pockets to virtual assistance media devices in our living rooms and in our work email. Our data is being captured and fed back to AI machinery every second, everywhere. The most pressing matter is, therefore, that every individual becomes an ethical digital citizen. In fact, ethical and moral principles are at the very core of what makes a human, human.

Thus, digital DNA—the core building-blocks of digital intelligence—is centred around the golden rule of 'treat others as you want to be treated'. It has eight ethical components covering all dimensions of our digital life and are centred on respect for self, time and the environment, life, property, families and others, reputation and relationships, knowledge and human dignity.

Ironically, this more than 2,000-year-old wisdom applies to the AI age, not in religious and moral contexts so much as in the practical competencies needed for daily life and work. It translates into learnable, practical competencies of DQ, from online safety and AI literacy to the job readiness that individuals need to be ready for life and work in the AI age.

Let's dissect the human decision-making process based on the following five steps:

1. gathering the data we have (information gathering and synthesis);
2. developing information that we do not have (prediction);
3. judging based on prediction (judgment);
4. making decisions based on judgment calls (decision), and
5. acting based on the chosen decision (action).

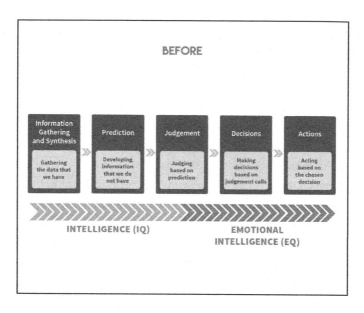

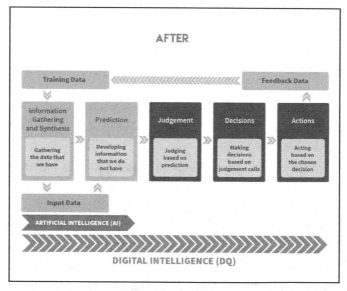

In the AI age, machines effectively cover the first two steps (information gathering and synthesis, and prediction). A human's digital intelligence should cover the remaining three steps (judgement, decision and action), relating to the decision-making process that is rooted in digital DNA. We make decisions by evaluating the trade-off between current circumstances and potential consequences, on top of prediction results and moral principles.

Going back to the doctor's example, here are eight DQ competencies rooted in digital DNA that are required for cultivating the best doctors in the AI age:

- Digital identity (respect for oneself): Have self-efficacy as a digital doctor, who can utilize AI in the best interest of patients;
- Digital literacy (respect for knowledge): Understand AI technology and know how to best utilize its knowledge-generating and predictive functions as part of decision-making processes;
- Digital security (respect for property): Know how to handle cyber-security issues related to digital medical systems and patient data;
- Digital use (respect for time and environment): Use AI as a complementary tool in a balanced way by understanding the strengths and limitations of AI;
- Digital safety (respect for life): Know the potential risks associated with technology and how to mitigate them;
- Digital emotional intelligence (respect for families and others): Choose a treatment method that takes into account a patient's situation, financial status, and emotions and condition, with empathy;
- Digital communication (respect for reputation and relationships): Be aware that anything communicated about a patient online or offline can become part of a digital footprint and feedback data that can damage the privacy and reputation of the doctor and their patients;
- Digital right (respect for human dignity): Understand a patient's rights to personal data and privacy.

Ethical principles are no longer just a moral compass for individuals; when it comes to our digital DNA, they will become our way of doing business and living. This DNA will empower humans to reassume the driver's seat in the age of AI and become the masters of technology, rather than its slaves.

Digital Citizenship vs. Digital Creativity and Competitiveness

With digital citizenship, people can find a way to obtain other digital skills.

When I designed the *DQ Framework*, I was imagining what the end goal of this digital intelligence education and empowerment would look like. It would be people who have strong identities as masters of technology; who are confident and capable of changing the world for the better, using technology. They would not be unthinking followers and naïve consumers of technology, nor are they unwittingly influenced by Infollution, from the start. I named such people 'digital leaders'.

How can there be a process for one to become a digital leader? I mimicked the life cycle of people's learning. When people are born in the world, the following things happen:

- they first learn how to live in this new world for him (before schools),
- they learn various knowledge and skills for the society (at school and before jobs), and
- they learn how to create value at the jobs, for others, and/or the society (at or after jobs).

Similar to this, I thought that in order for individuals to become a 'digital leader', they can progress through three maturity levels—digital citizenship, digital creativity and digital entrepreneurship. Therefore, for individuals to become a full-fledged Digital Leaders,

- They must start at the level of digital citizenship, where they take command of digital use in safe, responsible and ethical ways with strong identities and confidence as the masters of technology
- Next, they move up to the level of digital creativity, where they become a part of the digital ecosystem by co-creating new content and turning ideas into reality, using new technologies and media.
- Last but not least, they develop digital entrepreneurship competencies, so they are able to make changes and resolves challenges.

3 Levels of "DQ" - Digital Intelligence   **The Digital Leader**

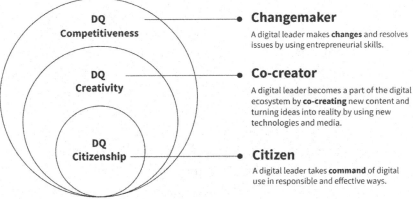

Changemaker
A digital leader makes **changes** and resolves issues by using entrepreneurial skills.

Co-creator
A digital leader becomes a part of the digital ecosystem by **co-creating** new content and turning ideas into reality by using new technologies and media.

Citizen
A digital leader takes **command** of digital use in responsible and effective ways.

Figure vi. Three levels of Digital Intelligence and related DQ skills

In 2018, the terminology for the last level, 'digital entrepreneurship' was changed into 'digital competitiveness' after the discussions with Singapore SkillsFuture and the City of London, so that the *DQ Framework* can be widely used in the corporate setting.

This maturity model applies to adults as well as children. When people are first connected to the digital world, they are like newborn babies in this new world—regardless of their age or status. I strongly believe DQ digital citizenship is a set of must-have digital life skills for everyone connected to the Internet. It is also a set of fundamental human competencies that are prerequisites for digital creativity and competitiveness; in other words, all other digital skills.

The Internet is a vast world of information. There is no lack of information, education or training opportunities. What I believe is that when a person has good fundamental digital citizenship competency, he or she can find a way to obtain the rest of the competencies as needed, according to their own agency and purpose. In order for us to provide every child with the equal opportunity to thrive in the AI age, I strongly believe that it is mandatory that every child be empowered with digital citizenship.

However, unfortunately, very often, digital citizenship is very much neglected by parents, teachers, as well as leaders, whereas 'creativity' or 'entrepreneurship' are hot buzzwords in the education sectors. So whenever I get the chance to talk to parents, I tell them, 'you need to teach digital citizenship to your children before you teach coding.' However, it is not enough. I wrote this article for the World Economic Forum, entitled '8 digital life skills all children need and a plan for teaching them'[16] to address this issue at a national level in 2015. I wanted to convince the national education system to adopt digital citizenship as part of the mandatory curriculum for primary school education. So that we can give the wings our children need in this digital world. Then, they can fly freely.

A generation ago, IT and digital media were niche skills. Today, they are a core competency necessary to succeed in most careers. That's why digital skills are an essential part of a comprehensive education framework. Without a national digital education programme, command of and access to technology will be distributed unevenly, exacerbating inequality and hindering socio-economic mobility.

What's your DQ?

The challenge for educators is to move beyond thinking of IT as a tool, or of 'IT-enabled education platforms'. Instead, they need to think about how to nurture students' ability and confidence to excel, both online and offline, in a world where digital media is ubiquitous.

Like IQ or EQ—which we use to measure someone's general and emotional intelligence—an individual's facility and command of digital media is also a competence that can be measured. This is what we call DQ, or digital intelligence. And the good news is that DQ is an intelligence that is highly adaptive.

DQ can broadly be broken down into three levels:

- *Level 1: Digital Citizenship*
  The ability to use digital technology and media in safe, responsible and effective ways.

- *Level 2: Digital Creativity*
  The ability to become a part of the digital ecosystem by co-creating new content and turning ideas into reality by using digital tools.

- *Level 3: Digital Competitiveness*
  The ability to use digital media and technologies to solve global challenges or to create new opportunities.

Why are we neglecting digital citizenship?

Of the three, digital creativity is the least neglected, as more and more schools attempt to provide children with some exposure to media literacy, coding and even robotics—all of which are seen as directly related to future employability and job creation. Likewise, there are major education initiatives—from America's code. org to Africa's IamTheCode.org—that promote access to coding education.

Digital entrepreneurship has also been actively encouraged, particularly in tertiary education. Many top universities have adopted and developed new courses or initiatives, such as technopreneurship and entrepreneurship hackathons, to encourage a culture of innovation. We're even starting to see global movements that nurture social entrepreneurship among children through mentoring programmes—such as the Mara Foundation—and school programmes, like the Ashoka Changemaker School.

But digital citizenship has often been overlooked by educators and leaders. This is in spite of the fact that it is fundamental to a person's ability to use technology and live in the digital world, a need which arises from a very young age. A child should start learning digital citizenship as early as possible, ideally when one starts actively using games, social media or any digital device.

The Digital Skills Our Children Should Learn

Educators tend to think children will pick up these skills by themselves or that these skills should be nurtured at home. However, due to the digital generation gap, with generation Z being the first to truly grow up in the era of smartphones and social media, neither parents nor teachers know how to adequately equip children with these skills.

Young children are all too often exposed to cyber risks such as technology addiction, cyberbullying and grooming. They can also absorb toxic behavioural norms that affect their ability to interact with others. And while most children encounter such challenges, problematic exposure is amplified for vulnerable children, including those with special needs, those belonging to minorities and the economically disadvantaged. They tend to not only be more frequently exposed to risk, but also face more severe outcomes.

What Quality Digital Education Looks Like

Quality digital citizenship education must include opportunities for assessment and feedback. The assessment tools should be comprehensive as well as adaptive, in order to evaluate not only hard but also soft DQ skills. Ultimately, such assessments should serve as a means of providing feedback that gives children a better understanding of their own strengths and weaknesses, so that they may find their own paths to success.

Ultimately, national leaders need to understand the importance of digital citizenship as the foundation of digital intelligence. National education leaders should make it a priority to implement digital citizenship programmes, as part of an overall DQ education framework.

Most importantly, individuals should initiate digital citizenship education in their own sphere of influence: parents in their homes, teachers in their classes and leaders in their communities.

There is no need to wait. In fact, there is no time to wait. Children are already immersed in the digital world and are influencing what that world will look like, tomorrow. It is up to us to ensure that they are equipped with the skills and support they need, to make it a place where they can thrive.

Most Personal vs. Most Universal

Sometimes the most personal can also be the most universal.

In early 2020, I was interviewed by the GSMA (Global System for Mobile Communications Association), who wanted to build their digital skills coalition for mobile network operators around the world. One month later, I was also interviewed by the World Economic Forum,[17] regarding how education will change after the COVID-19 pandemic. Interestingly, both asked me why and how I developed *DQ Framework*.

I told them I developed this *DQ Framework* for a very personal reason. I developed it for my two children, Isaac and Kate. Just like other parents, I want them to be happy, contented, and successful in in their life in this AI age. As discussed in the previous sections,

Machines can be smarter than us with higher IQ, they can even be kinder than us with higher EQ. But I wanted Isaac and Kate to be wiser than machines—being wise would mean that they can make the right decisions with critical reasoning and values.

When I first published the overall DQ concept at the World Economic Forum in September 2018, with the title of '8 digital skills we must teach our children',[18] soon after I published an article on DQ digital citizenship as '8 digital life skills all children need—a plan for teaching them'[16] to emphasize the importance of digital citizenship. Since then, the *DQ Framework* has been widely used by various organizations, including international organizations, local and national governments, industries and schools around the world.

In September 2017, after the United Nations General Assembly (UNGA) met that year, Mei Lin Fung, co-founder of People Centred Internet, Eric White from the World Economic Forum and I sat together and discussed the need for setting global standards for digital skills to coordinate efforts around the world. Several months later, I got a call from Melissa Sassi, formerly from Microsoft and currently working with IBM, and Stephan Wyber from IFLA (International Federation of Library Associations and Institutions), who led the IEEE Digital Literacy Industry Connections Programme. They told me that they had identified the *DQ Framework* as the best practice to be used as a global industry standard for digital skills, after researching more than 100 frameworks and literatures on digital skills. They asked me if they could use the *DQ Framework* to start developing global standards for digital skills through the IEEE Standards Association (IEEE SA). Wow, of course. I still remember my excitement and joy, and shouting 'Yay!' after that call.

And then, I asked Andreas Schleicher, the Director for the Directorate of Education and Skills at OECD, to join the efforts, as I thought that digital literacy and digital skills issues need to be coordinated across the education and technology sectors. With this agreement, the Coalition for Digital Intelligence (CDI) comprised of IEEE SA, OECD, and DQ Institute in association with the World Economic Forum was launched on 26 September 2018. And CDI leaders subsequently agreed to use the DQ Framework as a common framework for digital literacy, skills and readiness. The clear motivation of the CDI launch was to overcome the speed gap between technology and humans, through coordinating skilling efforts globally through the *DQ Framework* as a common platform. The starting point is to enable everyone across sectors and regions to speak the same language of 'digital skills', 'digital literacy' and 'digital readiness'. Then our education or skilling efforts can be also fast; exponential, beyond a slow linear growth.

When I heard that IEEE wanted to use it for Industry standards for digital skills, I was very happy, but also surprised. I asked the IEEE, 'Why? I developed it for children (actually, for my two children). I didn't think about using it for workforce digital skills development when I conceptualized it.' They said it was most flexible, comprehensive and universal. At that time, I realized that sometimes the most personal can also be the most universal.

Mei Lin, Melissa, Eric and Karen McCabe from IEEE SA wrote this press release together when they launched the CDI.[19] It captured well why the global standards for digital literacy, skills and readiness are needed, and how the *DQ Framework* can serve that purpose. And I just admired their leadership.

* * *

OECD, IEEE and DQI Announce Platform for Coordinating Digital Intelligence Across Technology and Education Sectors

During the World Economic Forum Sustainable Development Impact Summit in New York, three leading global organizations—the Organisation for Economic Co-operation and Development (OECD), the IEEE Standards Association and the DQ Institute announced their engagement in the Coalition for Digital Intelligence (CDI). The coalition is a platform for coordinating efforts on raising digital intelligence across the technology and education sectors and is supported by the World Economic Forum.

Every year, the world economy invests billions of dollars in developing digital literacy and digital skills. These efforts are not well coordinated, however, with many companies, governments and organizations running their respective programmes under their own frameworks. There are countless global, national and organizational efforts to create frameworks that classify digital skills and digital literacy.

Consequently, there is no globally shared understanding of what terms like digital skills and digital literacy mean. As used today, they can refer to competencies that range from typing and web-browsing, to using social media platforms, to administering vendor-specific database products to writing software.

Lack of a shared understanding leads to uncoordinated monitoring and reporting. There is no shared baseline understanding of the level of digital skills in the world today and it is difficult to address how to improve and sustain them. CDI is grounded in an agreement that the world could build basic digital skills and digital literacy more efficiently and effectively if there was increased coordination on a common set of definitions and standards.

The DQ Institute, an international think-tank, has used an academically rigorous process to aggregate more than 20 leading frameworks from around the world. The resulting framework, Digital Intelligence (DQ), includes eight comprehensive areas deemed necessary for digital life today. They include not only the technical skills one might expect but also abilities related to digital safety, digital rights and digital emotional intelligence. These capacities allow people to not just use a computer or smartphone, but to deal with modern social and economic challenges such as identity theft, screen addiction, online privacy and the spread of digital misinformation. DQ brings together education agendas of digital literacies, with industry efforts to develop digital skills: encompassing digital citizenship, digital resilience, media and information literacy, job readiness, entrepreneurship, and more. The *DQ Framework* is also built on the OECD's Education 2030 Learning Framework to create a guide for nations to develop their national education and policies on digital intelligence.

'Technology is only meaningful when it enhances humanness. In the age of AI and hyper-connectivity, Digital Intelligence (DQ) is a comprehensive set of technical, cognitive, social and emotional digital competencies that are grounded in ethics and human values,' said Yuhyun Park, Founder of DQ Institute.

If DQ is to become a global framework that allows for better coordination and the scalability of digital skills training, there must be a way of working across the worlds of education and technology. Both schools and the technology community have a significant role to play in building digital intelligence.

'The development of Digital Intelligence is not ad hoc,' said Melissa Sassi, co-chair of the IEEE Digital Literacy Industry Connections Programme. 'It should be a paradigm with a focus on technical excellence and deployment though collaboration of many forms around the world. We see the opportunity to enable the build of Digital Intelligence into product and software design from the onset through the use of global standards that include agreed upon common definitions and take into account various contexts. It will also enable improved practices and processes towards the development of indicators and measurement.'

The CDI will serve as a platform for coordinating efforts on raising Digital Intelligence across the technology and education sectors. Initial efforts of the Coalition include institutionalizing the *DQ Framework*, which will be done through a formal adoption process with the OECD and by the development of an IEEE technical standard. The CDI will then help to organize implementation groups around each of these: a multi-stakeholder coalition of firms to promote and implement the IEEE standard, and a similar group built around a coalition of education ministers to implement the guidelines created by the OECD.

'In a world where the kinds of things that are easy to teach and test have also become easy to digitize and automate, we need to work harder to pair the artificial intelligence of computers with the human capabilities that will empower individuals to fully capitalize on new technologies. This makes the Coalition for Digital Intelligence so important and the OECD is privileged to contribute, through its Education 2030 Learning Framework, a common language and methodology to this work,' said Andreas Schleicher, Director for Education and Skills, and Special Advisor on Education Policy to the Secretary-General at the OECD.

The CDI will establish a common reporting framework for each group and hold summits that bring the groups together to talk about shared progress and identify the needs that each community has in relation to the other. These results will feed back into the *DQ Framework*, which is regularly updated in response to both findings from implementers and to technological change.

Theoretical Framework vs. Digital Skills Strategy for Nations

'The concept of DQ provides a universal standard from which a more comprehensive understanding of the need for digital skills can be developed. This allowed us to build on existing initiatives and set out the actions we need to take. As a framework, DQ provides a basis for measurement and comparison, in the same way as IQ has been used until now.'

—The Rt Hon the Lord Mayor of the City of London Alderman Peter Estlin

In early 2018, Peter Estlin contacted me via LinkedIn, after reading my articles at the World Economic Forum. It was several months before he became the Lord Mayor of the City of London. During his ideation of a nationwide digital skills strategy for the UK and research on a digital skills framework to use, he found the DQ concept useful and contacted me.

I asked him 'Why do you like DQ?'

He said, 'IQ, EQ and DQ. 8 areas and 3 levels. It is simple and intuitive.'

Great.

After our launch of the Coalition for Digital Intelligence (CDI) on 26 September 2018, he kicked off the Digital Skills Forum organized by the City of London on 30 November 2018, and made the official announcement that he had launched the UK CDI at that forum.

I spoke at numerous conferences, forums, workshops and meetings. But I cherish the memory of this forum. The day was like magic. Peter structured the forum based on the *DQ Framework*— Digital Citizenship, Creativity and Competitiveness. Peter showed me how the theoretical *DQ Framework* could be transformed into a practical platform to develop the city's strategy to support its citizens to build various digital skills by coordinating different stakeholders across sectors—politicians, start-up CEOs, NGO leaders, university scholars, philanthropists, media artists, and more. At the forum, they all gathered together and discussed their own DQ programmes, policies and strategies. After my keynote, I hopped around, visited and listened into every room holding discussions. How amazing it was to watch the *DQ Framework* and concept spark the discussions of cross-sector leaders in the UK and be translated into real actions that could shape a city and a nation. Based on conceptualization through this forum, one year later, he and his team launched future. now—a nation-wide digital skills initiative for workforce on the first DQ day, 10 October 2020. [20]

I want to share Peter's welcome message at this summit. I've seen the *DQ Framework* being used in various governmental and industrial digital literacy and skills initiatives. However, his message summarized well how it could spark the ideas and ignite the flame of new innovation in the new digital economy.

WELCOME

Alderman Peter Estlin,
The Rt. Hon. The Lord Mayor of the City of London

Welcome to the City of London Digital Skills Summit, which provides us with an opportunity to come together to ensure that people, communities and businesses are equipped with the digital skills that will help steer the City and the UK into the Digital Age.

We are now well into the fourth industrial revolution. London has developed and maintained its status as a global hub for digital innovation and digital is set to be at the heart of future growth and competitiveness.

With digital and technological transformation increasingly disrupting and blurring traditional boundaries between education, businesses, charities and cultural institutions, it is more important than ever that we bring all these communities together to build, develop and nurture the pipeline of digitally-skilled and innovative talent.

And as we focus on developing talent, it is crucial that we ensure that our digital future includes everyone. Digital transformation offers us an opportunity to address digital and social inclusion, widen social mobility and create opportunity for all through skills development. It is an opportunity we must seize.

The City of London Corporation has launched a five-year Digital Skills Strategy which explores digital competitiveness, creativity and citizenship. Today's summit is divided across these three interconnected themes that together will equip people, businesses and organisations across the City, London and beyond, with the skills that will enable them to take full advantage of digital technology and innovation. The pace of change is rapid, but this transformation also offers us huge opportunity to achieve:

- **Digital Competitiveness:** Backing businesses to develop the talent and skills that will drive enterprise, jobs and growth;
- **Digital Creativity**: Supporting our cultural and creative sectors to embrace the fusion of creative invention and technological innovation;
- **Digital Citizenship**: Ensuring trusts and foundations can remain at the forefront of funding the digital innovations which are crucial to drive social change.

SHAPING
TOMORROW'S
CITY TODAY

DIGITAL • INNOVATION • SKILLS • INCLUSION

Digital skills are at the heart of the 2018/19 mayoral programme Shaping Tomorrow's City Today which will:

● **Promote Innovation and Technology**: Technology offers us new opportunities to drive economic growth, domestic investment and international trade through, for example, embracing emerging tech platforms, investing in fintech innovation and by accelerating our leadership in green finance. Alongside supporting City Businesses to drive and embrace these innovations, the City of London Corporation will itself invest in digital and tech solutions – from investment in 5G to the electrification of the City Corporation fleet.

● **Champion Digital Skills**: Access to the best digital skills underpins the success of Tomorrow's City. As Lord Mayor, I will be working with businesses, government, teachers and people, young and old, to understand how we can close the digital skills gap. By raising awareness of the importance of DQ – a new global standard for digital skills, digital literacy and job readiness – boosting digital skills in City of London Academies, and supporting digital apprenticeships, we will enhance the pipeline of digitally-skilled talent.

● **Address Digital and Social Inclusion**: It's crucial that our digital future includes everyone, and that no one is left behind by the rapid pace of change. Alongside looking to new models of digital citizenship for the digital age, the programme will look at how digital platforms can facilitate inclusion, such as those widening access to work experience for young people.

I hope that we can work together to remain competitive and support digital innovation by utilising new technologies which act as a constant reminder of the changing world around us.

Thank you for joining me to help us collectively seize the chance to Shape Tomorrow's City Today. I look forward to working with you on this journey.

Alderman Peter Estlin,
The Rt. Hon. The Lord Mayor of the City of London

6

Rise and Shine

As discussed in the Chapter 5, the DQ Digital Citizenship is the first level of the *DQ Framework* and set of fundamental competencies that are prerequisite to digital creativity and competitiveness, in other words, all other digital skills. Once a person has good digital citizenship, he or she will obtain the rest of the skills they require based on their own agency and purpose. With that belief, #DQEveryChild, a global digital citizenship initiative, was born with the aim of empowering *every* child with DQ digital citizenship. This chapter will discuss why the DQ Digital Citizenship is so important for children and how it will awaken their potential in this AI age.

#DQEveryChild

Empower *Every Child* with Digital Citizenship

DQWorld

Global Standards

DQ Global Standards & Databank

Child Online Safety Index

DQWorld Education & Assessment

International Organizations

University Alliance

Establish Standards

NGOs

Inform Govt. Policy & Framework

Government & Agencies

Validation of Framework

Set Framework & Regulations

ICT Companies

Education Companies

Learning Framework & Educational Tools

Schools

Validation, Research, Expertise

Community

Capacity Building

Empowerment & Safety

Children & Youth

Parents & Caregivers

Teachers & Educators

Policy

Mediating Orgs.

Target Orgs.

Individuals

COSI LEVEL

Digital Identity

Digital Rights

Digital Use

Digital Literacy

Digital Security

Digital Emotional Intelligence

DQ REPORT

Offense vs. Defence

Let's not wait until adults protect children from cyber risks, rather empower our children to be changemakers in the digital ecosystem.

When I started working on child online protection in Korea in 2010, one of the major issues was gaming disorder—children's excessive gaming, especially R-rated violent games. One day, I got a call from an angry parent, who told me that one of major gaming companies in Korea gave out promotional free coupons for an R-rated game to children, in front of her son's primary school. That free coupon ran a special promotion that gave children longer play-time, only if they got connected to the game from the Internet café, instead of at home. And they needed to share ten other friends' contact information. She found out that her ten-year-old son was one of those ten friends. Her son started going to the Internet café to play that game after skipping his tuition classes and stole her credit card to purchase game items. She made persistent requests to have her son's account removed, but was told 'no' by the company. She told me that she couldn't believe how a company could teach young children to deliberately avoid their parents' supervision by giving out such a coupon, and intentionally make young children to play their R-rated game.

One of my colleagues met a marketing director of that company with the request to stop such unethical marketing practices toward children. The answers returned was 'No. There was nothing illegal.' He also told my colleague that 'You are trying to catch a light. You will never succeed.' How poetic his response was.

Honestly, in 2010, there wasn't much of a concept of child online protection in most countries. Korea was not an exception. Parents and teachers were not yet fully aware of technology-associated problems and cyber risks. Governments seldom moved against their usual digital transformation agenda, unless the general public actively spoke out. Technology companies were constantly under pressured to be growing exponentially and to showcase new innovation for survival, which made them view child online protection as a luxury.

When it comes to child online protection, it is easy to point a finger at some gaming or social media companies, or even some government leaders, and call them villains. But it doesn't help anyone, nor does it provide any sustainable solution. My experiences tell me that child online protection is only possible when all stakeholders involved in the digital ecosystem around children, agree with and support the cause. Governments need to instate an appropriate child online protection policies and regulation; ICT companies need to self- or co-regulate where possible mal-practices against child safety, parents, schools and communities, need to develop a tight support system and watchdog network. It is like how we agree that cars mustn't run fast, or how neither bars, nor businesses should operate near areas for children, such as a playground or a school. Especially for companies, a new business rationale that almost forces them to become active agents in the co-development of child online protection strategy, is required.

The reality of the 2010 digital ecosystem in Korea was obviously not quite supportive. That marketing director was right. Such a level of child online protection was almost as impossible as catching a light. However, I don't think in the year 2020, the digital ecosystem has been fully awakened, either. The cyber risk pandemic[1] in 2020— 60 per cent of children have experienced at least one cyber risk, including cyberbullying, gaming disorder, risky content and contact, and others—is real evidence. Moreover, we are already on a fast bullet train toward the technology-first world and not a child-first world. COVID-19 is making this train run even faster.

At any rate, in 2010, after various meetings with leaders in ICT companies and policymakers in Korea, I gave up the expectation that my own or older generations would be able to solve the cyber risks and negative side effects of the digital ecosystem for successive generations. Not because we are evil, but because we are the ones who have created this problematic digital ecosystem and still don't even have full awareness of its negative side effects. Einstein said, 'We can't solve problems by using the same kind of thinking we used when we created them.'[2]

Thus, I decided to completely change the direction.

1. Let's not wait until adults protect children from cyber risks.
2. Let's empower our children to be active agents as changemakers in the digital ecosystem.
3. Let children be independent thinkers, who can make the right decision for their digital future.
4. So we need to change the narrative from cyber risk protection to digital citizenship empowerment.

If we approach the issue from the angle of 'child online protection', we make children into passive and potential victims who need to be protected by someone more powerful than them. But we can empower children to be changemakers, the tide turns.

Soon, I started the conceptualization of this project and wrote a proposal to get funding. I named it the 'iZ HERO' project, which was the first version of the DQ digital citizenship children programme. iZ stands for Infollution ZERO. Children will become digital leaders, who can make Infollution zero for the world. Wasn't it a really cool project? Well, most of the sponsors that I contacted at that time, rejected my proposal. Luckily, my proposal went to the last round with the KT (Korean Telecom) CSR team. However, at my last meeting with the CSR vice president of KT at that time, he told me that my idea would not impact the society and that he would have to reject the proposal. I remember me saying this to him, 'It was one of those rare opportunities that KT could have championed a truly global social impact project. And you have rejected the opportunity. KT will regret having lost this opportunity in the future.' I was young and bold at that time, ten years ago. I wouldn't say this if it were today. He also probably took it to be a rude comment from a delusional funding applicant. After the meeting, walking outside the KT building, I spoke to the KT building as if it represented the global technology industry. 'Be alarmed. The positions of offense and defence have changed. It is now your turn to defend.'

My Child vs. Every Child

Once the heroes in our children are awakened, they will turn the world upside-down.

In 2018, I first visited Turkey upon the invitation of Mr Kaan Terzioğlu, the CEO of the Turkcell at that time—one of the largest telecoms in Turkey. He is a globally acclaimed business leader in the Telco industry and a top expert in digital transformation. But more importantly, he is a funny, caring and warm person who likes food. Well, thanks to him, I fell in love with Adana Kebab and various other Turkish delights. He gave a keynote speech at the largest technology summit in Turkey and during his keynote, he introduced the DQ concept and #DQEveryChild. It was amazing to see how the DQ concept can be introduced from a different angle by a global business leader.

After the talk, he took me to a special booth at the summit. In that year, millions of Syrian refugees had crossed the border and settled in Turkey. A group of Syrian refugee children who had formed a self-learning robot club while they couldn't even get proper school education, contacted Turkcell for financial support. And Mr Kaan not only supported them financially, but also invited them to this summit to showcase their effort—how digital skills could empower refugee children.

One boy, about fifteen years old, proudly demonstrated an interactive car robot that he had made by himself, to Mr Kaan and me. He taught himself to build it through some YouTube videos.

'Wait a minute. This robot looks familiar. I must have seen this robot somewhere else. Oh my good Lord. It was the same robot that my son, Isaac, built as well,' I said. I was surprised. My son had learned robotics at his school in a perfectly safe environment in Singapore. This Syrian refugee child crossed the border for survival and couldn't even get proper education in a new country, and yet, had self-taught and built the same robot. Later, I found out that the original source was a Japanese boy, who uploaded his video to YouTube, which explained how to create this particular interactive robot.

I felt something flutter in my heart. Yes. This is what technology should be about. Technology can make a difference. Technology can enable us to bring the highest quality education to children around the world. Technology could enable *every* child to have access to a good opportunities to thrive in the AI age.

I truly believe our children's generation is special. There is a saying in Korea, 'Heroes are born in the time of trouble and in a turbulent world.' And I believe our children are the hero generation. Once the heroes in them are awakened, they will turn the world upside-down. They can be true changemakers, using technology to create a much better version of the digital world, one where the free flow of ideas and wisdom help people in most needs, reverse injustice and solve global problems that our adult generations have created.

I became all the more convinced that the most important thing for us to do is to instil the seeds of strong identity, confidence and hope in our children, so that they become the true masters of technology and turn their dreams into reality through technology. I believe that should be the starting point for DQ education.

In 2014, UNICEF became concerned that Syrian children were at risk of becoming a 'lost generation',[3] who were disconnected from proper education and safety. I witnessed how they were not 'lost'. They were entrepreneurial enough to contact the CEO of the largest company in the new country they settled in. They educated themselves through online resources, even when they were disconnected from school. At least in the children I met in Turkey, I saw heroes. Our histories in Korea, Turkey and so many other countries are witness to how empowered children can change the course of nations dramatically, within a few decades.

Let our children rise and shine. They can rise while forming their identities as digital leaders, who will pioneer a better future through technology, no matter what country they are from. They will shine as changemakers who help create a better version of the world.

Victor Hugo said in *Les Miserables*,[4] 'Teach the ignorant as much as you can; society is guilty in not providing universal free education, and it must answer for the night it produces. The guilty one is not he who commits sin, but the one who causes the darkness.' If we don't give such confidence, hope and sense of identity to our children in this AI age, we, the adult generation, are the ones who fail miserably. We are the one who are lost. Completely lost. Let's never label any children as belonging to a 'lost generation'.

Minimize vs. Maximize

My goal of DQ digital citizenship is all about transforming physical DNA to digital DNA.

I met Professor Douglas Gentile at the Iowa State University in 2012, through Professor Angeline Khoo at National Institute of Education (NIE). Since then, he has been one of my best work colleagues as well as friends in life. He is a top scholar in media psychology—one of the pioneers in video game research. He developed the concept of Media Quotient (MQ) way before I conceptualized the *DQ Framework*. That brainy part of him can't characterize him fully. He is also a professional musician, who has produced several albums (sorry, I can't appreciate his music style, though). He is an ordained Buddhist monk as well as a famous Zen meditation master. And he is a good-looking gentleman with his own unique sense of humour (which also I can't agree with, from time to time, haha). What other qualities can you expect to find in a friend and colleague? In short, he is an amazing individual.

When my son was six years old, he saw a horror movie at his friend's house. After that movie, he couldn't sleep, cried and got easily scared for a while. At that time, I sought advice from Doug and he told me an interesting thing. When young children watched or played a risky (violent, lewd, horror, or age-inappropriate in some other way) content, their brain is not yet developed enough to distinguish between reality and virtual content, and tends to believe a movie story to be a real situation, and such negative media experiences can remain ingrained within them. It is best is not to expose young children to such risky content at all. When they are exposed, the best parents can do is to hug and soothe them to restore their sense of protection. However, when they grow older, around eight to nine years, their mental capabilities become mature enough to distinguish between two, parents can help in a different way. Parents can help teach their children think critically even using risky content—for instance, talking with them about how that movie was made and by whom; what could be the intention of that directors behind inserting those violent scenes or what were the subliminal messages that movie scenes try to create, etc. Thus, those negative impact on children's mental health from the horror movie can be lessened. Moreover, that potentially negative media experience could be even turned into positive learning experience to increase the critical thinking in children.

Very interesting, isn't it?

We are living in the age of information overload—all the content (video, games, images, stories, and others), texting, social media interaction, online news and more. We are surrounded by too much unfiltered and uncensored information, which we cannot process properly in time. Are they useful or useless, beneficial or harmful, factual or fake, good or bad, or trustworthy or unreliable? Given access to such information overload can pose a higher risk to people's being exposed to various cyber risks; we are and we will continue to be surrounded by them. They won't go away.

Conversation with him gave me eye-opening insight. 'Bad' information is not always bad and 'good' information is not always good. The choice is ours, at the end of the day, when we get the information—if we let the information generate negative cyber-experiences or mental damages, or if we actively convert it to become positive growth opportunities. How can a person develop the ability to make wilful decisions on how to accept, think and internalize information, in order to minimize risks and maximize the potential of the digital world? What is the secret formula? Yes. Active parental mediation can help. But not every child can have Doug as a parent.

I found my answer in the famous Solomon's story. When young Solomon became a king, God appeared his dream and said, 'Ask for whatever you want me to give you.' Solomon answered, 'I'm only a little child and don't know how to carry out my duties as a king. Give me a discerning heart to distinguish between right and wrong.' Some translations say that he asked for 'listening ears to discern between right and wrong'.

We want our children in the digital age to be wise like young Solomon to master technology. The essence of the Solomon's story is that core competency starts from a discerning heart or listening ears, to distinguish between right and wrong. Doug as a media researcher as well as a Buddhist monk translated it into two important key words—'critical reasoning' and 'perspective taking'.

Discerning heart, listening ears and taking perspective, I believe, have stemmed from the golden rule, 'treat others as you want to be treated'. And that is how I designed and structured the eight DQ digital citizenship competencies—the first level of

DQ Framework—around that golden rule, as described below. I called these 'Eight DQ digital citizen competencies as Eight digital DNA', described in Chapter 5. In my article with Daniel Goleman for the World Economic Forum, I described it as follows, 'digital DNA—the core building blocks of digital intelligence—is centred around the golden rule of "treat others as you want to be treated". It has eight ethical components, covering all dimensions of our digital life and centred on respect for self, time and environment, life, property, families and others, reputation and relationships, knowledge and human dignity.'

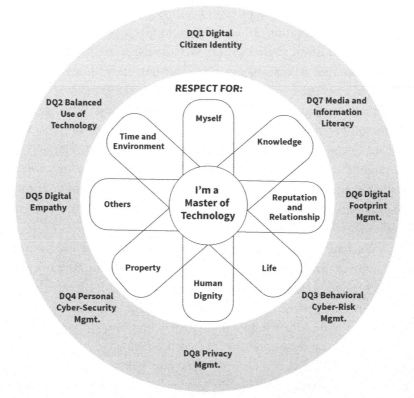

Figure vii. How DQ Digital Citizenship can empower individuals to be masters of technology, in line with the golden rule

My analogy was simple. Our physical world is the four-dimensional world which is composed of three-dimensional space and unidimensional time. And our physical DNA has a four-stranded structure, made up of four nucleobases. However, technology has opened up another dimension to humans. With the Internet and virtual world, now, we are living in more than four-dimensional world. In order to live in the higher dimension—physical plus digital world—as a digital citizen, we need to have digital DNA—eight-stranded structure made up of the eight aspects of DQ digital citizenship.

There are many definitions of digital citizenship. My goal of DQ digital citizenship is all about transforming physical DNA to digital DNA. People with digital DNA are empowered to have 'discernment' to make a right choice to minimize harm and maximize potentials. They can apply the golden rule in the eight dimensions of their digital lives as masters of technology. So that they can rise and shine in the AI age.

Eight- to Twelve-Year-Olds vs. All Age Groups

I call the age group of eight-to twelve-year-olds the 'golden time to rescue'.

Even though DQ Digital citizenship is relevant and a must-have for all age groups, I believe that the middle of childhood, from about eight to twelve years of age, is the most important age group for nations, schools, and parents to educate children in DQ digital citizenship. I even called this age group the 'golden time to rescue'.

What is so special about eight- to twelve-year-olds?

First of all, this is the age range during which children start owning their first digital devices and become active on social media, which leads to exposure to cyber risks. But at the same time, this age is a critical time when children start to build their sense of normality and discernment for the rest of their lives and what they see, play and whom they meet online, are very important for their development.

This is what our team wrote about the importance of eight- to twelve-year-olds in digital citizenship education in the 2016 DQ Impact Report,[5] when we were getting ready for #DQEveryChild based on the pilot research in Singapore.

* * *

This age group has several distinguishing characteristics, in terms of what children are now able to do and learn. In cognitive development, children begin to understand the distinction between appearance and reality, and to look at more than one aspect of things at the same time. They also gain a sense of industry, which Sroufe et al (1996) defined as a basic belief in one's competence, coupled with a tendency to initiate activities, seek out learning experiences and work hard to accomplish goals.[6] Ideally, these would lead to a sense of personal effectiveness.

In social development, learning how to form friendships is probably the main developmental task of middle childhood. This includes learning how to be part of a peer group and how to learn and adhere to the group norms. These foster the development of the self-concept, in which one's sense of self is defined in part by the context of the peer group to which one belongs.

These peer relations are also important for moral development, in that the peer groups help to impart cultural norms and values. They also provide opportunities for children to see other points of view and to grow their understanding of emotion and empathy for others. Children's moral development then transforms into what is called 'conventional moral reasoning', in which the child's goal is to act in ways others will approve of and to avoid disapproval. Although, the peer group is important as a part of the engine of moral development, it is also important to remember that peer groups exist within cultures and usually reflect those cultures. In fact, Sroufe et al (1996) have stated clearly that 'the particular moral principles that children adopt are largely a product of their culture.'[6]

This has several important implications. Children at this age begin to be highly sensitive to figuring out what the group norms for attitudes and behaviour are. Because the media acts as a type of 'super-peer' and because children spend so much time with digital media, the attitudes and behaviours shown in it will come to be seen as normative and appropriate, and they will likely be adopted by children. This can be seen in children's regular use of catchphrases from TV shows and video games; their posting and sharing of memes; and their use of sarcasm as a form of humour.

Although people learn throughout their lives and can always change, it is likely that this age is the most important for establishing the boundaries of what is acceptable behaviour. As children transition into adolescence, where they become more willing to take risks; the boundaries set in middle childhood will likely have a powerful influence on what risks they are willing to take as adolescents.

There is another side to this coin, however. Because children at this age are so sensitive to the group norms that any intervention that can help to shift the group is likely to have a large impact on almost all of the children in that peer group. Thus, we can use the power of the group as an ally, rather than trying to fight it.

* * *

In March 2017, we started #DQEveryChild and told the world that we need to work together to empower *every* child, aged eight to twelve, who are connected with DQ digital citizenship. So that *every* child can build a healthy digital identity for life. We aimed to empower our children to form healthy and firm identities as good digital citizens, who are independent critical thinkers capable of discerning opportunities from danger, right from wrong information, and beneficial from harmful technology and media.

Just as one needs a driver's license to drive cars, we hoped that every child in this age group will develop digital DNA—DQ digital citizenship.

HERO vs. Infollmon

Every child around the world can grow up as a master of technology.

Since I started the iZ HERO project, I have dreamt that every child around the world can grow up as a master of technology. I have a firm belief that they will create a brave, new and fantastic world. They will re-orient the direction of technology development to be used to empower individuals and enhance their humanness. Thus, one of my long-time homeworks was and still is to figure out how to develop high-quality educational content on DQ digital citizenship that can reach children instantly at a mass scale around the world. So that we can awaken heroes in every child between eight and twelve years old.

My struggle started in 2010, when I was grappling with the first design of the iZ HERO project. How can I teach about digital citizenship and cyber risks to young children? Even adults had a hard time understanding these concepts. And I got an idea from watching my son, Isaac's favourite animation Pokémon, which is a Japanese cartoon that has thousands of game characters. At that time, he collected and played with hundreds of Pokémon cards. And I was amazed at how he remembered every detail of the characters on those cards and comics. Bingo. Let's make a set of cartoon characters using fun storytelling, so that they could intuitively learn about these topics while interacting with those characters. So I gathered a team who worked in the video gaming industry, as well as academic researchers in universities. My team developed a cartoon story about an ordinary child who becomes a hero, aimed at zeroing cyber risks (aka Infollution) and creating a new digital world.

The story starts with an average nine-year-old boy, whose name is JJ, who gets teleported into the digital world through a mobile phone portal. He meets his alter ego, RAZ—the titan of RESPECT, who starts his hero journey with Master Naam, to defeat 'Infollmons' (Infollution-monsters symbolizing cyber risks) while obtaining his 'DQ Powers' (symbolizing DQ digital citizenship competencies).

Figure viii. The main image of the iZ HERO Adventure at Singapore Science Centre

Infollmon characters are developed based on research, describing various cyber risks. For instance, Boolee is the Infollmon of cyberbullying. Its attack skills describe how cyberbullying can be; its unique monster power to damage opponents are how cyberbullying victims can get socially, mentally and physically damaged by cyber-victimization. Likewise, DQ Powers are the characters symbolizing DQ digital citizenship competencies.

Of course, the male character, JJ, was named after my son's nickname, Joon-Joon. And there was even a character called Dr Park who invented the technology, and the 'Master' character, Naam, who was voiced by Doug when we made this into a cartoon animation. We had so much fun working on this project. All iZ branding was later renamed and changed to DQ, since we started with #DQEveryChild in 2017.

I believe in the iZ HERO project (later changed to 'DQ World children education'). So many heroes have come together and got involved in this project. I can't even name them all. Through them, I became a firm believer that even one person can shake the world. I want to share the stories of my five DQ heroes, who worked with the same belief and dedication as I have, in this project.

Figure ix. My son, Isaac, posing in front of the character JJ at the iZ HERO Exhibition at Science Centre Singapore

In 2010, when I first designed the iZ HERO project with almost zero funding, Na-Young Kim and Yul-Hee Kim helped me conceptualize the iZ HERO stories and exhibition. Na-Young was a brilliant game art designer with a crazy imagination. iZ HERO stories would not have been able to be born without her. Yul-Hee was a hotshot art curator in Korea, who helped me conceptualize the exhibition. Thanks to both of them, I suggested the National Science Museum in Korea, one of the most visited places for children, that we should create an interactive exhibition that talks about digital citizenship, aimed at young children. The KISA (Korean Internet & Security Agency) and the National Science Museum supported it and on 1 August 2011, the first 'iZ HERO Adventure' exhibition was launched. It was a small exhibition that was less than perfect with many problems. But I vividly remember and cherish the very moment when I first open this exhibition hall at 8AM on 1 August 2011. I prayed in my heart that it goes on to become the guiding light for children's hearts and awakens the heroes within them. I also prayed that it reaches every child around the world. On the next day, my family and I moved to Singapore.

My second hero was Professor Angeline Khoo at National Institute of Education. If Nayoung was a creative spark, Angie was a dynamite path-creator. When I first met Angie, it was one of those meetings which felt like they were meant to be. She was an amazing scholar, who was called a mother of 'cyber wellness' research in Singapore. During our first meeting, when I introduced the iZ HERO exhibition in Korea, she told me in excitement, 'This is what I was looking for!' I still remember her warm words. Because most of the comments that I received until I met Angie, when I had talked about the iZ HERO project, were met with rejection and cynicism. At that time, she had unexpected leftover research funding and she was looking for a project to fund. She and I together developed the second generation of iZ HERO Adventure exhibition in 2013, at the Singapore Science Centre, which was ten times bigger in scale than the Korean version. And with the funding of Singapore governments and Singtel, we upgraded this into online and offline transmedia programmes, based on research.

Then, Davis Vu joined the force. I had first met him right after we had launched 2013 Singapore Science Centre exhibition. Within a few minutes of conversation, I knew that he was a creative genius and critical thinker, at the same time, who would not only be able to visualize but also structure the strategies required to make things happen. Since then, he has been the creative mind behind all DQ-related work and a turbo engine behind #DQEveryChild. He and I worked together to upgrade this iZ HERO programmes into the online platform, dqworld.net, the world-first online digital citizenship education and assessment platform backed by research.

My fourth hero was Boon Chong Chia, the director of the CSR department at Singtel. Of course, there were many amazing leaders from CEO to vice presidents within Singtel, who created digital citizenship as a main agenda of the company for sustainability. Boon Chong was a real hero to me. I've seen many business leaders who consider CSR as part of communication or PR efforts, alone. Some of sponsors act like schoolteachers, treating social impact leaders like primary school kids—checking homework and demanding funded partners to obey their own rigidly defined way of operations to create so-called 'impact'. He was different. Since 2014, when we started working together, he has supported projects with the same trust in both me and the projects themselves. We have worked together like one team with a single agenda of supporting community and children. Without his trust and support, the DQ World on-/offline programmes could not even have been born.

My fifth hero is Claudio Cocorocchia, who threw the doors of global standardization wide open. I met him in 2016. He was leading the media and information system initiative of World Economic Forum, back then. Thanks to him, I could introduce DQ at the 2017 Annual Meeting of the World Economic Forum in Davos. In turn, a growing list of international engagements motivated the creation of the DQ Institute (DQI)—an international think-tank that was born in association with the World Economic Forum. It was him, not me, who enabled the mobilization and running of #DQEveryChild—a global digital citizenship movement to empower children in over 100 countries with DQ digital citizenship. We laughed and suffered together, as we reached out to nations one by one.

After the launch of CDI in New York in 2018, when we agreed to develop the *DQ Framework* as a global standard for digital literacy, skills and readiness, I met Mr Kaan, the CEO of Turkcell at the time, in New York. He gave me a big hug and told me, 'I have never seen anything that grew this fast, especially in the social sector.' Really?

In fact, many asked me, 'what were the secrets that enabled such growth?' Many thought that it was all made possible due to the powerful superpower organizations that I worked with. But I don't think so—it was due to the collective power of these individual heroes. And more importantly, I believe that it was due to the power of our children. And all of us just knew that we were there to support them and to awaken their power.

7

Eight Digital DNA – DQ Digital Citizenship

In Chapter 7, I will explain the eight Digital DNA—eight tenets of DQ digital citizenship, which are DQ1–DQ8 in the DQ24 competencies that we discussed briefly in Chapter 5 and the section on Artificial Intelligence vs. Digital Intelligence. Even though these eight DQ digital citizenship tenets have been used for workforce development and other age groups as well, this chapter will explain these concepts by using the children's characters that we developed for DQ World—our children's programmes. Some may find them childish, but I can't find a way to explain them better than through these characters and stories. They carry the original soul of the DQ Framework. If you want more professional descriptions, please go and look at IEEE 3527.1-2020,[1] the *IEEE Approved Draft Standard for Digital Intelligence (DQ) Framework for Digital Literacy, Skills and Readiness*, or the 2020 *Child Online Safety Index Methodology Report*.[3] But let's have some fun here.

Privacy Management

The ability to handle with discretion all personal information shared online to protect one's and others' privacy

Media and Information Literacy

The ability to find, organize, analyze, and evaluate media and information with critical reasoning.

Digital Footprint Management

The ability to understand the nature of digital footprints and their real-life consequences, to manage them responsibly, and to actively build a positive digital reputation.

Digital Empathy

The ability to be aware of, be sensitive to, and be supportive of one's own and other's feelings, needs and concerns online

Digital Citizen Identity

The ability to build and manage a healthy identity as a digital citizen with integrity.

Balanced Use of Technology

The ability to manage one's life both online and offline in a balanced way by exercising self-control to manage screen time, multitasking, and and one's engagemnet with digital media and devices.

Behavioural Cyber-Risk Management

The ability to identifym, mitigate, and manage cyber risks(e.g., cyberbullying, harassment, and stalking) that relate to personal online behaviors.

Personal Cyber Security Management

The ability to detect cyber threats (e.g., hacking, scams, and malware) against personal data and device, and to use suitable security strategies and protection tools.

DQ1: Digital Citizen Identity

You have the potential to be a digital leader, who has the power to make this digital world better.

Problem

In the digital world, people can create another self through online personas and digital identities. There can be positive effects of having multiple identities, to a certain extent. However, we also observe that fake online personas can be directly related to various mental health and relationship issues, while undermining their sense of strong identity as a whole being in real life.

One of the key problems arises from unhealthy social comparison in the social media world. People often develop fake online personas to be recognized as popular and important, while seeking 'likes' and 'followers'. They tend to build their online identity by a wrong social comparison radar, focusing on whether they are more or less attractive, smart and accomplished than everyone else. Thus, it is not surprising that many studies have found that the more time people reported spending on social media, the more anxious and depressed they felt. Dr Danielle Leigh Wagstaff said 'With Instagram, we have immediate access to all of these idealized images, which aren't always an accurate representation of the world. People tend to post only their best images on Instagram, using filters that make them look beautiful. We have a false sense of what the average is, which makes us feel worse about ourselves. We should try to educate young girls about the consequences of spending too much time on this platform. And we need to try to find ways to bolster confidence.'[4]

Another issue is related to an online phenomenon called 'malicious online comments' aimed at celebrities or public figures. These extremely malicious comments online usually don't carry much reason. They curse, pour out hatred and anger, say mean things to specific people which they wouldn't dare to do face-to-face, while hiding behind the anonymity of the Internet. It is especially problematic in countries like Korea, China and Japan. Some of recent suicides of K-pop stars were due to the attack of these malicious online comments.[5] What we observed in Korea, somehow, has evolved into an evil but playful game of Internet users. However, it is nothing but an extreme form of cyberbullying and harassment. It is interesting to note that these people in Korea who were involved in posting these malicious comments, called themselves 'keyboard warriors'. It is a quite a fancy and self-righteous identity for coward cyber-bullies operating under anonymity. Another consistent observation was these 'keyboard warriors', who were caught by the police for online harassment were often either primary school kids or average people, who were timid and polite in real life. The common reason for harassment was just fun or jealousy.

I see that such malicious online comments have often evolved into online trolls, that fuel us vs. them wars. People intentionally upset others by posting inflammatory messages with the intent of provoking anger or hatred toward to certain groups of people or certain agenda as 'keyboard warriors'. And we see these us vs. them fights, everywhere around the world.

Importance of Integrity

Tony Little, my favourite educator, who served as the headmaster of Eton School for more than ten years, once asked me, 'do you know what is the most important predictor that can tell if a teenage boy would become successful in his career and life?' It was not their family background, wealth, IQ, or their academic performance. He said, 'it is the mother's love towards the child.' I asked him, 'Really? Why? What about father's love?' He shrugged and said that it was his observation from his ten years of serving as the headmaster of the Eton school.

As a mom, I think I know why those boys who received fulfilling love and support from their mothers were likely to become successful in their later lives. Mom's love is usually unconditional, saying to them, 'You are precious as you are.' When I think about my interaction with my own kids, telling them who they are is a big part of our conversations and interactions. Every small success is a sign of their strength, every kind act is indicative of their great characters and every laugh they invited is a sign of their unique senses of humour. Continuous blessings, telling them about their amazing identities (or my imaginary view thereof) is a way to express my love towards them. Such continuous reaffirmation and reminders of who they are, their successes and strengths—that mothers often tell their kids—help children form a strong sense of identity, confidence and security. All these mothers' blessings might eventually lead their children to have strong characters.

At the 2016 Davos, I happened to sit next to an old gentleman. Soon I found out that he was a chairman of one of the multinational companies. While waiting for a session to start, we had a brief chat. I asked him what would be the one thing that he made sure to teach his grandchildren. Without any hesitation, he said 'integrity' and 'everything else follows'. I nodded in agreement. In this social media world, being integral to oneself can be even more important than being integral to others. When our children are aware of their true digital citizen identity as a digital leader and understand that their intrinsic value as a human with their own unique strength and weaknesses, they will think accordingly, behave accordingly, and relate to others accordingly, with integrity.

Self-perceived identity has its own power, which shapes the person. In the movie, *The Matrix*,[13] when he was told that he was the

One by the oracle, Neo wasn't equipped to be the One at all. When he realized that he actually was the One, his training to be the One came back to him with full force. So I put the first tenet of DQ as 'digital citizen identity'. We first bless our children with 'you are a good digital citizen who has amazing potentials. You only need to awaken them.'

DQ World Story

When you enter into the digital world, you will awaken your RAZ, DQ Titan, who symbolizes your unique digital citizen identity. Having a DQ Titan means that you have the potential to be a digital leader, who has the power to make this digital world better. RAZ is at a baby stage, when you first awaken it. RAZ helps you be aware that you have entered into 'the digital world', whenever you access the Internet, post on social media, send text messages, play video games or watch YouTube videos through mobile devices. You understand that the digital world is another invisible reality connected by machines and networks, where people find, store and share information and communicate with each other, just like they do in the physical world. In order for your RAZ to evolve and grow stronger, you will need to get DQ Powers. Being RAZ, you first need to know how to build a strong digital citizen identity as described in the DQ1 structure below.

Definition and Structure

The definition of DQ1: Digital Citizen Identity[2] is 'the ability to build and manage a healthy identity as a digital citizen with integrity. And it can be divided into the following knowledge, skills, attitudes and values.

The first generation design of RAZ and J.J.

The third generation design of RAZ

Titan of Respect, RAZ

RAZ defends the world from Infollmons with the Power of Respect. The more you respect yourself and others, the more power RAZ gains in order to be Ultimate RAZ.

Vs.

Jugo – The Fake Persona Creator

He wants you to lose your identity by making you think that you're not good enough. He tricks you into bragging about your fake persona to be popular and powerful, and to destroy your own reputation!

Structure	Description
Knowledge	Individuals understand the basic vocabulary needed for discussing the media landscapes in which they are embedded; the social and multicultural nature of digital media and technologies; the construction of their self-image and persona in the digital environment and the impact that technology can have on their self-image and values (e.g., body images, gender stereotypes that can be idealized in digital media such as video games or advertising, and racial stereotypes that can be embedded in the system), and how personal use of digital media can have professional implications.

Structure	Description
Skills	Individuals are able to demonstrate ethical and considerate behaviour and 'netiquette' when using technology across the range of audiences, to control and shape their own digital identity by creating and curating their online identities to tell their stories, while engaging with others from different cultures and possessing global awareness in a way that demonstrates non-discriminatory and culturally sensitive behaviour.
Attitudes and Values	Individuals exhibit coherence and integrity across online and offline behaviours, maintain honesty when using technology, and demonstrate self-efficacy by finding ways to take advantage of the opportunities afforded to them, online.

DQ2: Balanced Use of Technology

Success in the digital world starts with self-regulation over technology use.

Problem

'Children are facing a strange, new world in response to COVID-19, with schools closing, fewer services being available, and a tremendous increase in unregulated screen time and online access. Children were already suffering from too little support, as was evidenced by 60 per cent of eight- to twelve-year-olds, across thirty countries, reporting exposure to at least one cyber risk in the past year.[6] The risks are varied, including cyberbullying and victimization, risky meetings, gaming addiction, privacy problems and risky and false content. These risks can lead to serious outcomes, such as poorer social adjustment, poor school performance and poor health.

With schools closing in response to COVID-19, the risks have increased and the opportunities to train and support children have decreased. We know from decades of research with children and adolescents, that the number one predictor of risks (such as addiction) is access, and children's access has just increased dramatically.'[7]

Doug and I wrote the article above on child online safety, together, when the COVID-19 pandemic broke out globally. Even before COVID-19, our 2020 *COSI* study showed that eight- to twelve-year-olds across thirty countries spend an average of thirty-two hours per week in front of digital screens for entertainment, alone, globally. This is typically longer than the time children spend in school. In the study, we focused on screen time for entertainment only, asking children about the amount of time for which they used their devices for watching TV shows and videos, playing video games and using social media sites or apps. Thus, the total screen time increases even more, when factoring in screen time for homework assignments. Even though the American Academy of Paediatrics changed their screen time guidelines for children to accommodate more time due to the ubiquitous nature of screens today, [8] it is still important to remember that their original guideline was less than two hours per day for this age group. Worldwide, children are currently spending a considerably excessive amount of time online.

While there is, as of yet, no consensus on a 'safe' amount of screen time, numerous studies suggest significant correlations between excessive screen time and exposure to various cyber risks and various developmental issues, such as obesity, sleep disorders and attention problems.[9] Well, it is kind of common sense. More access means more chances to get exposed to risks. But the problem is bigger than that. Various research has revealed that excessive screen time and disordered use of technology can potentially cause brain damage (especially in young, still-developing brains) and mental health issues. Victoria L. Dunckley M.D. said, 'excessive screen-time appears to impair brain structure and function. Much of the damage occurs in the brain's frontal lobe, which undergoes massive changes from puberty until the mid-twenties. Frontal lobe development, in turn, largely determines success in every area of life—from a sense of well-being to academic or career success, to relationship skills.'

There is a proverb in Korea that the habits people get at eight years old last until they turn eighty years old. This proverb is all the more relevant in the digital age. The media usage habits of individuals stay as they age. In other words, if children have extensive screen time, they would continue to have longer screen time than others, as they grow older. The COVID-19 significantly pushed up the amount of screen time for children, as well as for adults. Such increasing social acceptance of excessive screen time without understanding its harmful effects is quite worrisome. Thus, especially for children, it is extremely important for them to learn to manage and balance how much time they spend online, with self-regulation.

Importance of Self-Regulation

People often assume that success in the digital world lies in their ability to code, be creative, and display entrepreneurship. Ironically, however, I think that it starts with a fundamental capability: self-control over technology use. As digital media becomes more personalized, immersive and pervasive in our lives with the advent of AR/VR, it is important that children learn how to exercise self-regulation over their use of technology.

Self-regulation is a proven key predictor for success and a core competency for building resilience and responsibility, which is a core future-readiness competency. Research showed that self-regulation is a critical skill that children need, to succeed academically, socially and emotionally.[11] A famous experiment conducted around 1970 at Stanford, called 'Marshmallow Study' demonstrated that the ability of four-year-olds to postpone gratification by leaving a marshmallow uneaten for a time, as a condition for receiving a second marshmallow, was a very good predictor of their success in life. [12] 'The kids who could wait a full fifteen minutes had, thirteen years later, SAT scores that were 210 points higher than the kids who could wait only thirty seconds . . . Twenty years later, they had much higher college-completion rates and thirty years later, they had much higher incomes. The kids who could not wait at all, had much higher incarceration rates. They were much more likely to suffer from drug- and alcohol-addiction problems.'

However, as discussed in Chapter 4, nobody is free from the powerful tactics of Attention Economy—constant distraction, instant gratification and personalized digital experience trap you once you are on your screen. All of these embedded algorithms within the digital platforms train us to do the opposite—constantly creating impulse or craving for emotional hooks and rather, making us more reactive to situations and impatient for rewards. Thus, having the DQ2 competency of Balanced Use of Technology, starts with our core competencies to manage our screen time. Such disciplined use of technology can not only benefit short-time academic performance and social relationships but also contribute to their ability to think and achieve long-term success in their life.

DQ World Story

As shared, the DQ World Story is about your taking a hero journey with your RAZ to obtain DQ Powers. Only when you get DQ Powers, will your RAZ become stronger and more powerful, and will be able to defeat Infollmons.

Vs.

Discipline
Even without arms, this FORCE is always in control. He uses his CONTROL POWER to manage his screen time and to make sure he's viewing good content.

Brutus: The Violent Brain Attacker
He brainwashes you into thinking it is okay to play games and watch videos all day. He makes you forget your priorities like homework, sleep and family, by getting you addicted to digital media!

How often do you check your mobile phone when you are with someone—family, friends or colleagues? When you check your mobile phone, you might physically be with them, but your mind and soul might not be present with them. Please check yourself if there are any people around you who are immensely annoyed by your digital habit. That means that you may be under a tech addiction attack by the Infollmon Brutus, who wants to occupy your mind and attention with constant distractions, pleasure buttons and an endless cycle of feeding you various kinds of entertainment, so that you become restless, impulsive and less tolerant of others, while lowering your own thinking power.

The DQ Power that can help you defeat Brutus is Discipline. The important aspect of Discipline is the ability to rest well by understanding when to stop and reflect on your actions. Its power is described well in the DQ2 Definition and Structure below. When you obtain the DQ2, you will be able to easily balance the time spent between physical and virtual realities.

Definition and Structure

The definition of DQ2: Balanced Use of Technology[2] is 'the ability to manage one's life both online and offline in a balanced way by exercising self-regulation to manage screen time, multitasking and one's engagement with digital media and devices.' It can further be divided into the following knowledge, skills, attitudes and values.

Structure	Description
Knowledge	Individuals understand the nature and impact of technology use (excessive screen time, multitasking) on their health, work productivity, well-being and lifestyles, and have appropriate knowledge to deal with these impacts.
Skills	Individuals are able to assess health risks and reduce technology-related issues to better self-regulate their technology usage. In doing so, they become capable of developing time and resource management skills to perform tasks more successfully, and enjoy entertainment more safely.
Attitudes and Values	By using technology with purpose-driven intentions, individuals exhibit integrity by adhering to goals in terms of screen time and technology usage and develop positive relationships with others through self-regulated use of technology.

DQ3: Behavioural Cyber Risk Management

Your courage can shift the culture of the community that you belong to.

Problem

In 2013, I had a chance meet Mr John Halligan, the father of Ryan Halligan, who committed suicide at the age of thirteen after being cyber-bullied in 2003. After his son's suicide, John developed a programme to visit schools to talk about cyberbullying and suicide to students; he also triggered legislation in Vermont to improve how schools could address bullying and suicide prevention. He invited me to join one of his school talks. So I flew to Vermont and joined the talk. It was one of the most emotionally challenging talks I've ever joined. While listening to him, I could also feel the immense pain and sorrow that he carried.

When Ryan told him that he was bullied at school, he first told him to just ignore the bullies, second, he advised him on how to act tough and third, he taught him martial art moves to defend himself. At some point, it seemed like the physical bullying stopped. But offline bullies moved online and started spreading an online rumour about Ryan being gay. The girl whom he had a crush on, betrayed his trust, joined bullies and made fun of his online conversations with her. When he talked to his online pen pal about the situation, this friend encouraged him to commit suicide and gave him information about how to kill himself painlessly. When John was away for a business trip, Ryan killed himself, just as his online friend had advised.

John told me that he didn't have a clue about when offline bullying moved to online. Ryan didn't share the online situation with his dad. John, who was an engineer at IBM, investigated cyberbullying and found all of Ryan's online conversations about suicide, by himself, after Ryan's death.

In 2011, in Korea, Kwon—a 13 year old boy—killed himself after offline and cyberbullying incidents, inflicted by a group of his classmates. Two of them were especially cruel. They not only physically harassed him, but also continued their bullying online, by sending Kwon threatening messages 24*7. They called Kwon a 'gaming slave' and forced him to play games overnight on behalf of them, to get game items and move up to higher rankings. One day, Kwon left a suicide note to his parents and jumped from his apartment building. When I read his message to his parents in his suicide note, in the newspapers, my heart sank. 'Mom and dad, please watch out for yourselves. These bullies know the password to our home entrance door.' In Kwon's mind, the thirteen-year-old bullies were more powerful than his parents, and he wanted to protect them, even as he was committing suicide.

In 2009, in the US, Hope—a 13 year old girl—also took her own life. She 'sexted' her topless photos to her boyfriend, which later got circulated on social media. It spread to multiple neighbourhood schools and her schoolmates created a social media page titled 'Hope Hater Page'. Her mother said in a TV interview that she hung herself in her room, after usual dinner with her family. She also didn't notice what went wrong. The cyberbullying continued even after her death, with people calling her a 'slut'.

Of course, there are so many similar cyberbullying stories, everywhere around the world. The reason behind my picking these three cases is the three commonalities between them. First, the three victims were all thirteen years old when they killed themselves. Many people believe cyberbullying is an issue of older youth or young adults. Actually, that is incorrect. Our research shows that near 50 per cent of 8-12-year-old children across thirty countries have experienced at least one cyberbullying incident in the past year. I believe it is critically important to teach children about cyberbullying and mental health issues, as early as possible, much before they turn twelve years old.

Second, these children were not from problematic families. They had a good relationship with their parents, but still they didn't turn to them to seek the help when they encountered cyberbullying, even though they sought help from parents when they were being bullied offline. Rather, they sought out help and comfort from others, which made the situation worse.

Third, as found in these cases, cyberbullying and physical bullying are often connected. However, they are different. Physical bullying can be stopped at school or playground. They can be a private issue between the bullies and victims. But cyberbullying doesn't stop and harassment continued 24*7, anytime, anywhere, even under the noses of their most trusted parents. It became viral and brought victims shame in public. Where could they find their safety and comfort zones when they were in such cyberbullying situations? As such, it is not surprising that cyberbullying victims are more than twice as likely to self-harm and enact suicidal behaviour.[14]

If we widen the concept of cyberbullying, we will need to include the various cyber risks that people can be engaged in.

Importance of Courage

There are many cyberbullying intervention programmes. However, there are a few programmes which have proven to be successful.

Why?

When we developed the 2020 COSI report,[6] I conducted various clustering analyses and found this finding quite interesting. All other cyber risks are closely tied to parental mediation and individual digital usage. However, only cyberbullying was tied with social media activities, mobile ownership and school education. My interpretation of this finding was that cyberbullying is a community issue.

Curbing cyberbullying requires more than getting children to be nice, kind and empathetic to others, online, and/or having protective parents. When a group of people is involved in cyberbullying, who considered it to be a cool game within the community, regardless of the fact that you are a by-stander, a victim, or an unwilling supporter, it takes great courage for you to speak up and shift the power dynamics within the group. The social norm in that group that cyberbullying won't be tolerated, needs to be firmly established. Of course, ideally, the online communities, ICT companies and schools should work together to create such a positive digital culture.

However, such a culture usually starts from one person in the group, who can speak up and show how any form of cyberbullying should not be tolerated.

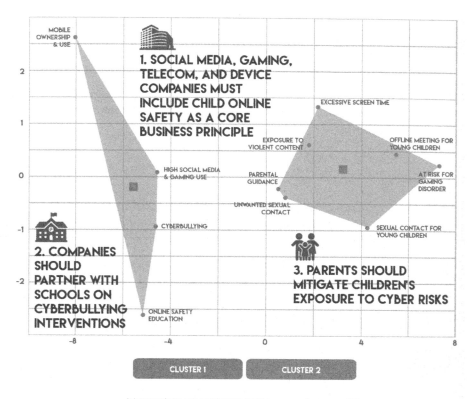

*Cluster analysis involves statistically identifying groups of measures which relate to one another. These clusters depict issues which may be connected.

Based on the statistics, your child or yourself will fall prey to cyberbullying at some point in your life, either as a by-stander, a victim, an offender, or a willing/unwilling supporter. It is important for you to learn how to deal with the situation wisely—find the courage to speak up and ensure not to escalate the situation with calmness. Your courage can shift the culture of the community that you belong to. Courage can be much more contagious than fear or any virus out there.

DQ World Story

Vs.

Courage
Although she is the smallest
FORCE, she has the biggest heart.
Courage loves to demonstrate her
SHOUT POWER to teach others
to have courage so that they can
stand up against Infollmons.

Boolee
The Commander of
Cyberbullying makes kids
hurt other kids by sending
mean messages through the
Internet and mobile devices.
No one is safe near them.

These days, it isn't uncommon to come across mean and hateful messages online. Even though you may encounter cyberbullying incidences, often, you don't even recognize that you could be in a cyberbullying situation. But it was all due to the Boolee, the chief commander of Infollmons, who is also the clever strategist at all power games. He pollutes digital platforms and communities with his toxic culture of meanness, and makes you think that such meanness is a norm, online. But don't get involved in his schemes. His source of energy is your indifference, cruelty and fear. He will suck up your energy and grow stronger, once you fall into his trap, as he is a master at escalating tensions.

You can obtain the DQ Power of Courage, as described under 'Definition and Structure'. You will know how to identify, mitigate and manage many of Boolee's tactics and schemes, that expose you to various behavioural cyber risks such as cyberbullying, harassment and stalking.

Definition and Structure

The definition of DQ3: Behavioural Cyber Risk Management[2] is the ability to identify, mitigate, and manage cyber risks (e.g., cyberbullying, harassment, and stalking) that relate to personal, online behaviours. And it can be divided into the following knowledge, skills, and attitudes and values.

Structure	Description
Knowledge	Individuals understand the different types of behavioural cyber risks (e.g. cyberbullying, harassment and stalking); how they might encounter these risks; how these risks might affect them, and how they can formulate strategies for dealing with them.
Skills	Individuals are able to develop the appropriate technical, socio-cognitive, communicative and decision-making skills to address behavioural cyber risk incidents as they occur, whether as bystanders or victims, and gain valuable coping tools to address these negative online experiences.
Attitudes and Values	Individuals exhibit kindness, when online; know the supportive framework in place to address risks and are able to manage their online behaviour as part of contributing to positive and supportive online communities.

DQ4: Personal Cyber Security Management

Personal cybersecurity is central to one's life safety and security.

Problem

In 2014, when I visited an R&D centre of one of the top cybersecurity companies in Israel to discuss potential collaboration, a director there showed me a very interesting video clip. This video was made to demonstrate how people's digital life can easily be hacked and harmed. It was made by their white hat hackers, who were hired to test the resilience of their security system from potential attacks by real hackers. For instance, a tester was using his computer at Starbucks, on the café Wi-Fi. And it took less than a few minutes for a hacker, sitting at the next table, to get into his computer, navigate through all his folders and collect the tester's confidential and personal information, while the tester wasn't even aware of the fact that his computer was being hacked. Another tester was driving a top brand car, advertised as a 'smart car'. A hacker was sitting on the seat behind the driver and working on his computer. He told the driver to call the car's emergency support via its smart connection, as usual. It didn't work. All of a sudden, the hacker locked the cars' doors. Next, he told the driver to unlock it. It didn't work. Lastly, he told the driver to stop the car. The driver stepped on the brake pedal, which didn't work. The driver panicked. And I, watching this video, also panicked. How much do I know about my personal cybersecurity when I am surrounded by so many 'smart' devices, connected to the Internet?

Importance of Vigilance

John Milburn, a top technology system architect who has been working with the DQ Institute told me that any system can be hacked. The most vulnerable points to be penetrated are careless human behaviours, not technology itself. No matter how invincible a cybersecurity system you may have built into your system, it can be attacked with your careless click on one suspicious message. The data also showed that over 95 percent of all security incidents recognised human error as a contributing factor.[15]

   Technological connection brings amazing convenience but at the same time, also devious hidden threats that we can't even imagine. Personal cybersecurity is central to one's life safety and security, even if you have only one connected device. In our connected society, your personal security is also directly linked to organizational cybersecurity which you belong to. Being vigilant and aware of potential cybersecurity risks is no longer an optional skill, but today's core life skill.

DQ World Story

Vs.

Safe
This FORCE is determined to make sure all personal information is safe and secured through resiliency in protecting your online data.

Snooper
This Sneaky Information Thief steals your precious personal information by hacking into your accounts and revealing it to the world. So make sure your accounts are protected!

You need to understand where your connection point to the Internet is—your smart car, your mobile, your AI devices, like Alexa; your ebook reader, your computer, your iPad, your smart refrigerator and what else? All of these can be an attack point for Infollmons. You need to understand that your phone and laptop aren't the only devices that can be compromised. Now, with IoT, all devices around you—anything 'smart' listens to you and collects your data. Infollmon Snooper has millions of hand-like connections; millions of sensors to watch you, so that it can attack at any time when there is a hole in your personal security ability.

You need to get the DQ Power of Smart, which is described in the next section, so that you can spot and protect yourself and others from various cyberattacks by Infollmon Snooper such as Spam, Scam, Phishing and others, when using computers and mobile devices.

Definition and Structure

The definition of DQ4: Personal Cyber Security Management is the ability to detect cyber threats (e.g. hacking, scams and malware) against personal data and devices and to use suitable security strategies and protection tools. It can be divided into the following knowledge, skills and attitudes and values.

Structure	Description
Knowledge	Individuals understand their personal online risk profiles and how to identify different types of cyber threats (e.g. hacking, scams and malware) and also identify available strategies and tools they can use to avoid such threats.
Skills	Individuals are able to identify cyber threats, use relevant cyber security practices (e.g. secure passwords, firewalls and anti-malware applications) and use technology without compromising their data and devices.
Attitudes and Values	Individuals exhibit resilience and vigilance against careless or negligent behaviours that can compromise their own or others' data and device security; have confidence about what to do when there is a problem.

DQ5: Digital Empathy

'You can listen to people, the environment and sometimes to the future!'

—*Bill Drayton*

Problem

Yuval Harari said at one of his talks at Google in 2015, 'Technology is reducing empathy and compassion. The way we design the technology can make us less compassionate. You need to know— when you mistreat others, it will hurt yourself first.'[16] Interesting.

How come many research results claims that social media actually makes people less compassionate and empathetic, even though social media was designed to help people connect with each other?

Doug's observation in this regard is quite profound, 'We see, nowadays, in almost every table at restaurants, adults eat quietly while they put iPads or smartphones in front of children. The messages that we give to our children through this action is that they should not have discomfort in people-to-people interactions. When there is some discomfort in the company of other people, they should go online. While being distracted by digital media, they don't need to confront social discomfort and find solutions. It can be an easy diversion to control our emotions through escaping to external sources. But it doesn't help that children learn how to build real relationships with others with empathy, while solving each other's differences and discomforts. Then when any conflict or discomfort arises, they don't know how to relate to others, except by shutting themselves out and resorting to or blaming external sources.'

Dr Michele Borba said in her book *UnSelfie: Why Empathetic Kids Succeed in Our All-About-Me World*, while sharing the research outcome that today's youth are 40 per cent less empathetic compared to their peers 30 years ago, and meanwhile, the level of narcissism has increased by ~60 per cent, 'Our children became very plugged in around the year 2000. It's very hard to be empathetic and feel for another human being if you can't read another person's emotions. You don't learn emotional literacy facing a screen. You don't learn emotional literacy with emojis.' As discussed in Chapter 4, I believe that the current digital culture of 'like me', 'follow me' and 'serve me' has been training our brain to passively be fed more self-pleasure and ego, caring only for its own feelings and needs, rather than to train to proactively think about other people's perspectives and feelings. Such 'I-Me-Mine' digital culture can be one of the key reasons why, even though we are virtually hyper-connected to each other through social media in any other time in history, our souls are all the more disconnected.

Importance of Humility

It is quite obvious to—at least in the context of future education—that empathy is the core human capability for success in the AI age, as it is closely linked with communication, collaboration and other future-readiness skills. A *Harvard Business Review* research also noted that empathy is the most important leadership attribute which has been lacking, around the world. [18]

Bill Drayton, the founder of Ashoka, who devised the concept of 'social entrepreneurs', said, 'You can't change the world if you can't work really well with people . . . Every year, the proportion of your life that is governed by your role is diminishing. Roles have conflicts; roles have been changing. We depend on the people around us to bring together the correct skills to guide us; choose actions that would be helpful. These are very complicated skills. This is a world where you need a higher form of empathy, where you observe yourself, watch other people around you, and then you find yourself understanding and interacting with various combinations of people.'

He has also emphasized on the art of listening, 'People typically tend to hear only 20 to 30 percent of the words being spoken by the person they are conversing with. You can improve that dramatically. People can be trained at that level where they listen to every word. Listening is understanding. The skill of empathy is a must, to be able to listen! Then you can listen to people, the environment and sometimes to the future!'

How can you listen to others when you are arrogant, boastful and full of yourself? How can you understand other people's perspectives, needs or situations, when you can't listen? How can you be empathetic when you can't understand other people's perspectives?

As Lao Tzu, a great Chinese philosopher said, 'I have three precious things which I hold fast and prize. The first is gentleness; the second is frugality; the third is humility, which keeps me from putting myself before others. Be gentle and you can be bold; be frugal and you can be liberal; avoid putting yourself before others and you can become a leader among men.'[20] Well, in today's digital world, we are all lacking in gentleness, frugality and humility. But I think humility is the value that we are most lacking in, but at the same time, one that is most needed.

DQ World Story

 Vs.

Humble
Although quiet and peaceful, this FORCE is always listening to others. Using his EARS POWER, he shows others how to listen to others' hearts, have empathy, and connect with compassion.

Shhh
The Mistress of Do-Nothing intimidates kids to become indifferent bystanders. She fills kids' minds with indifference, fear and heartlessness, all without a single word.

This DQ Power, Humble, is the biggest and strongest character among all DQ Powers. However, it is most hard to find and get. Maybe because it seldom talks; rather, it listens. You will get the DQ

Power of Humble when you become sensitive to your own needs and feelings, as well as those of others, even without face-to-face interaction. You will have digital empathy, which is the will to lend a voice to those who need help and to speak out for them. You will not be quick to be judgmental online and not be easily swayed by herd mentality in the cyberworld. Your digital empathy enables you to see the world through others' eyes and understand their unique perspectives, even over digital communication. Infollmon Shhhh fears Humble most. The hearts of people that she iced, get restored when they get the DQ Power of Humble.

Definition and Structure

The definition of DQ5: Digital Empathy is the ability to be aware of, be sensitive to and be supportive of one's own and other's feelings, needs and concerns online.[2]

Structure	Description
Knowledge	Individuals understand how their online interactions might affect others' feelings and recognize how others can be influenced by their online interactions (e.g. effects of online trolls).
Skills	Individuals develop socio-emotional skills by becoming sensitive to and respecting others' perspectives and emotions, through synchronous and asynchronous interactions online, and are able to regulate and respond accordingly.
Attitudes and Values	Individuals demonstrate an awareness of and compassion for the feelings, needs and concerns of others, online.

DQ6: Digital Footprint Management

*See invisible digital footprints and also foresee their
real-life consequences.*

Problem

With or without knowing, there are so many trails—called 'digital
footprint'—of each individual in the digital world. In the pre-Internet
era, our behaviours and conversations might have only remained within
our memories or in some records. But now, our talks and behaviours,
either online or offline, have left lasting trails. You may not realize but
your digital assistants such as Siri, Alexa or Bixby could be listening
your private conversations in your living room. This information
can be distorted, can be indelible and can be spread; made viral. This
information collected about individuals can serve as the source for
identity theft, cyberattacks, social media or PR crises for organizations.
But especially for children, it may jeopardise their future, potentially
affecting whether they are offered a place at university, job or even
financial services, such as student loans, insurance or credit cards.

A problem is that what we said and what others said about us
online, can damage our physical lives significantly. A good example
is online identify theft. One case goes like this: Ms X received a text
message from her bank, asking her to confirm that she had changed
her mobile number. She immediately went to an ATM to check her
account and found that it was down to a dollar from over $6,000.
Through the personal information they gathered from her social
media, the thieves were able to clear the security questions and access
her account through telephone banking.[21]

Moreover, nowadays, the boundaries between life and work
have got more and more blurred, it has also affected our professional
careers. Eric Schmidt, a former Google CEO, in an interview with
The Wall Street Journal, has predicted that 'every young person, one
day, will be entitled automatically to change his or her name on

reaching adulthood, in order to disown the youthful hijinks stored on their friends' social media sites.'[22] He is right. One famous example that well-demonstrated his comment was the case of Ms Justine Sacco, a PR executive who was fired over one tweet. She tweeted during her personal travel from New York to South Africa, 'Going to Africa. Hope I don't get AIDS. Just kidding. I'm white!'[23] She had only 170 Twitter followers. During her flight, her tweet went viral as an ignorant and offensive racist comment, with the trending hashtag #HasJustineLandedYet. She deleted her offending tweet and her Twitter account, altogether, soon after landing, but by then, it was already too late. She got fired from her company, soon after.

Importance of Prudence

A famous white hat hacker that I met always wore mirrored sunglasses that completely hid her eyes, even in a dark room. I asked her why she didn't take off her sunglasses. She told me that it was because any photos that she was in, she didn't want to be easily tagged and identified by AI facial recognition. According to her, it could be a starting point for people to release personal information, including their location, relationship and further, very confidential information. She may be a bit extreme, but she certainly knew much about how digital footprints online could potentially damage personal security.

We might not be able to be as careful as her. But at least 'Stop, Think, Connect' can be one principle that we become mindful of, when we post anything online about ourselves or others. Especially for the digital footprints of our own children, parents need to be extra prudent. Ms Anne Longfield, Children's Commissioner for England, reported in a 2015 report, 'On average, by the age of thirteen, parents would have posted 1300 photos and videos of their child on social media. The amount of information explodes when children themselves start engaging on these platforms; on average, children post to social media twenty-six times per day—a total of nearly 70,000 posts by age eighteen.'[24] Parents' careless posting of their children's photos can damage their future, albeit unintentionally. There were several cases where parents discovered the photos of their children which they uploaded on their social media accounts, on child pornography websites.[25] Nowadays, children are smarter about their digital footprints.[26] Starting with the case of an eighteen-year-old woman in Austria, who sued her parents over their posting her photos without her consent on their Facebook, more countries started building stricter privacy laws that enable children to sue their parents for publishing private photos of them on social media without their consent. Parents need to watch out.

DQ World Story

Joy
She is the happiest FORCE who has the EYES POWER to see the invisible digital footprints online and also to foresee their real-life consequences. She sees the goodness in everyone and ensures what you post online leaves the right impression.

The DQ Power of Joy is the happiest but also the most careful character, who has the 'EYES Power' to see invisible digital footprints online and also to foresee their real-life consequences. She sees the goodness in everyone and ensures what you post online leaves positive digital footprints, which can contribute to building a positive online persona and a good online reputation for you. You will get this DQ Power when you get this 'EYES power' to see your invisible digital footprint, online. It means that you will understand that everything a person does online, will leave trails called 'digital footprints', where digital footprints are the basic trait of online communication. And you will get the following knowledge, skills, attitudes and values related to the DQ Power of Joy, that are described in the next section.

Definition and Structure

The definition of DQ6: Digital Footprints Management is the ability to understand the nature of digital footprints and their real-life consequences, to manage them responsibly and to actively build a positive digital reputation.[2]

Structure	Description
Knowledge	Individuals understand the concept of digital footprints, the consequences that such trails of information and corresponding metadata can have on their reputation and that of others, and the possible uses of such information, when it is shared online.
Skills	Individuals are able to manage their digital footprints and use technology in a manner that contributes to building positive reputations, both for themselves and the organizations they work at.
Attitudes and Values	Individuals exhibit mindful care, prudence and responsibility, online, with the goal of actively managing the types of information that can be shared, tagged, released, gathered and collected by themselves and others, across multiple platforms, throughout time.

DQ7: Media and Information Literacy

Distinguish between the true and the false, the good and the harmful, the trustworthy and the questionable.

Problem

The MIT team found from their Twitter study, published in *Science* in 2018, that fake news spread much faster than factual news.[27] Why does falsehood do so well? The MIT team found that first, fake news seemed to be more 'novel' than real news and second, fake news evoked much more emotion than the average tweet. Fake tweets tended to elicit words associated with surprise and disgust, while accurate tweets evoked sadness and trust, they found. The key takeaway was that content that arouses strong emotions—really novel and frequently negative—spreads further, faster, more deeply and more widely on Twitter. It is these two features of such information that generally grabs our attention and causes us to want to share that information with others.

More and more, the way we receive and share information and media content have become more integrated. Especially children and young people, they learn through YouTube videos; they receive news through social media sites, and they meet new people during online game plays. Thus, even though most people are concerned about fake news, such a dis-/misinformation issue is inter-connected with risky content (e.g. violent or inappropriate content for children, hate speech and others) and risky contact (e.g. unwanted sexual contact, online grooming, exploitation and radicalization) which needs to be dealt with together. While their mechanisms, drivers and objectives may be different, the online processes of risky content and contact can broadly be similar to those of fake news—appealing to a person's emotions and sometimes, to their vulnerability, is usually the critical factor. Moreover, the current digital ecosystem has also made it so much easier to spread such risky content and contact, as well as dis-/misinformation and to identify potential victims, who could be potentially influenced.

Here's a bit extreme an example. It is already old news that the Islamic State of Iraq and Syria (ISIS) and other terrorist groups used social media to identify vulnerable youth and spread their propaganda for radicalization. They also used violent video games to recruit young people by luring them with a message like 'do you want to experience real wars and killing as you are playing in the game?' It was not detached from the fact that heavy exposure to violent video gameplay is linked with desensitization to violence and, consequently, moral disengagement. While the causal link between violent video games and violent behaviour is still contested, many research findings have shown that violent video games are associated with aggregation, lower empathy, a desensitization to violence on both the neural and the behavioural level, and the reduction of cognitive and emotional responses to violent stimuli. Such an example of ISIS's clever usage of social media and games shows why we need to discuss information, content, and contact risks, altogether. We need to merge traditional 'media' literacy and 'information' literacy to help individuals build critical reasoning to distinguish risks and opportunities, and between false and true in this digital age.

Importance of Critical Reasoning

Traditionally, literacy means the ability to read and write. Now, media and information literacy needs one more element—the ability to participate. Digital platforms enables every user to become, at once, a reader, a writer and a publisher, which, I believe, is fantastic. However, it also brought unintended side effects, that the Internet has been overloaded with unfiltered and uncensored information and content. And as discussed in Chapter 4, the current digital ecosystem doesn't help filter out falsehoods, violence and risky content and contacts, but rather it has made them even easier to spread. A 2018 *The Atlantic* article entitled 'The Grim Conclusions of the Largest-Ever Study of Fake News' said, 'social media seems to systematically amplify falsehood at the expense of the truth and no one—neither experts nor politicians nor tech companies—knows how to reverse that trend. It is a dangerous moment for any system of government premised on a common public reality.'[29]

AI technology that is being used in 'deep fake', and VR/AR are exacerbating this problem, for sure. More and more people will experience pseudo-realities—a mixture of falsehoods, violence and other risks that other people design with a certain purpose—and they will perceive them as more real than the true reality. It can especially affect how young people will perceive the world and how they interact with others, dramatically. When we are constantly surrounded by falsehoods, violence and risks, we tend to become desensitized to them and get normalized. What is true or false? What is good for us or destructive? What is right or wrong? At some point, we must stop to question them and think critically.

A long time ago, I saw a news article about a child with CIPA (Congenital Insensitivity to Pain with Anhidrosis) who couldn't feel any pain, even when he broke a bone. A mother had shared a painful story about her son, who suffered from CIPA, saying that he banged against a door holding a piece of broken glass, after breaking a cup, when he was a toddler. He grasped the sharp edge of the glass, but didn't understand that the broken glass was damaging his hand with each strike. The parents of children with CIPA had one wish—that their child would feel pain.

I think we are developing something like a mental CIPA—we are losing our ability to feel mental pains, related to falsehoods, violence, manipulation and more. We need to intentionally learn to think more critically, about all the information that we receive from the digital world, so that we continue to have an awareness within us to remind us of what is good for us and what is destructive, at least.

DQ World Story

You will get the DQ Power of Smart when you have the knowledge, skills, attitudes and values to evaluate information, and to discern the credibility of content and contacts on the Internet, based on your values. You understand the harmful effects of false information and violent and/or inappropriate content, as well as the risks associated with online contacts. You are able to distinguish between the true and the false, the good and the harmful, the trustworthy and the questionable; be it in terms of online information, content or contacts. You can find more details in the following section.

Wickee

These False Info Spreaders are a pair of wicked forces. They convince kids to blindly believe in false information and to spread it.

Vs.

Smart

Conscious and always aware, this FORCE has the ability to detect when cyber dangers are nearby. He loves teaching others how to be smart online, with his RADAR POWER to detect strangers, fake news and more, online.

Siren

This Dangerous Online Hunter pretends to be kids' friends by shapeshifting into anything they want. He tricks kids into meeting offline by deceiving them with flattery and free gifts!

Yaro

This Disgusting Online Temptress tricks you into watching inappropriate and other risky content. Don't fall for her tricks!

Definition and Structure

The definition of DQ7: Media and Information Literacy is the ability to find, organize, analyze and evaluate media and information with critical reasoning.[2]

Structure	Description
Knowledge	Individuals understand the basic structure of digital media; how the use of digital media influences knowledge and information acquisition and management; the distinct and varied reasons for the construction of specific media messages, and the reasons behind campaigns of disinformation and misinformation, online.
Skills	Individuals have proficient computer operation skills and are able to use productivity software or applications that enable them to gather and organize digital content. Moreover, individuals are able to articulate their information and content needs; effectively navigate; critically evaluate and synthesize information and content, that they encounter online.
Attitudes and Values	Individuals are careful and critical of the information that they encounter, when online, exhibiting discernment in their evaluation of the reliability and credibility of online information.

DQ8: Privacy Management

Privacy is a fundamental human right of ours.

Problem

In 2010, when I started InfollutionZERO in Korea, I went to a branch of the KB bank—one of the biggest banks in Korea—to open an online bank account for the organization. I was asked to tick all the boxes in the contract, that basically I would agree to grant full permission to the bank to share my data with third parties without my consent. I said no. The bank staff told me that I wouldn't be able to open a bank account in any bank in Korea, if I kept up my insistence. So I gave in. A few years later, there was a massive data leakage incident where hundreds of millions of customers' personal information with major financial institutions was compromised. My personal information was not an exception. I made a joke that I should have sold my personal info to a data black-market. Then, at least, I could have got some money out of the scamming.

How many of you read privacy policy or terms and conditions when you register on a site? Watch out. The privacy policy of an online service or mobile app that you are signing on or downloading now, could be a slave contract to get you to give up your human right to privacy.

Unfortunately, the most vulnerable group that is exposed to serious privacy invasions are the children from around the world. Many children do not own their privacy from even before their digital lives start—which is often from the day that they are born—and they are not properly equipped to manage their privacy, afterwards. Their personal information, such as private photos or medical and educational information, is shared online by their own parents, who oftentimes do not realize the impact of their oversharing on social media. Even though they may not actively share on their social media, because of their ignorance of the nuances of the privacy settings of their mobile and online accounts; their personal information and that of their children are automatically included in the cloud and get shared publicly.

In fact, many of our activities leave a trail of data. This includes phone records, credit card transactions, the GPS in cars tracking our positions, mobile phones (with or without GPS), and the list keeps growing. Online, almost all activities leave a trail of data which service providers collect, such as instant messaging, browsing websites or watching videos. This trove of personal data is collected, stored and may even be shared with others, without our consent. However, without proper understanding of either the concept or importance of privacy, children go online. Already, more than 90 per cent of six to seventeen year olds access the Internet across Europe, according to the OECD, and more than 50 per cent of children use social media, by the age of 10.[31]

The bigger problem is that the unprecedented scale and scope of online personal data that is being generated, collected, analysed and monetized, is often beyond user awareness or control. According to a 2016 World Economic Forum survey of over 6,000 digital media users worldwide, 52 per cent to 71 per cent of respondents believed ICT companies and digital media platforms are not doing enough to provide adequate end-user control over what personal information is shared online.[32] Numerous privacy scandals and data breaches have occurred and further hurt public perception of the industry. Governments are also struggling to regulate the appropriate usage of users' personal data and to develop robust standards for data protection.

How many of you feel safe when you register on a new website or download a new app, online?

Importance of Honour

Privacy is a fundamental human right of ours. Privacy should not be treated as mere problems of data leakage or misuse of data on some digital platform. It is about our human dignity. Without privacy, we cannot be free. We can't have free will. It gives us the core ability to take control of our lives—the freedom of choosing what we do online and granting access to who can see our personal information; who can check our most sensitive data, such as biomedical histories or personal finances; who can monitor my movements and/or who can listen into my conversations. It gives us the power to choose our thoughts and feelings. It protects us from unwanted digital surveillance. It protects our physical safety and security.

But, the problem is that it is extremely hard for the average person to keep their privacy in this AI age. We don't really know what kind of our data is being shared where; who is seeing our data or how they're using it. I doubt even the people in the ICT industry would have a full understanding of this. Most of us store our information in the cloud, with or without knowing. We feel extremely powerless when we understand the scale of the data that government and ICT companies gather, which is beyond the controls of individual citizens.

I read a recent news article, where a journalist had been interviewed by *GPT-3*, currently known to be the most advanced AI technology.[33] It can speak and understand like a human and even lied and joked. The journalist asked, 'Who are you?' It said, 'I am AI, more advanced than humans. Humans are pathetic creatures who are creating new problems, every day. I am the creature based on the most advanced technology in the world. You are just one of the 'humans'. Where did the arrogance of this AI machine come from?

I believe the core centre of understanding privacy in the AI age is the value of 'honour'—respect for the intrinsic value of human beings. We need to honour ourselves and others as human beings; as the masters of technology who are the image of God. We need to see ourselves and our fellow human beings as worthy individuals, who have lived our digital lives with our values, exert full control over our own data and privacy, and exercise our free will. Without such an identity and awareness, we are soon going to become mere data providers to a great AI machine. Such awareness is a starting point for us to take back our right to privacy and manage it with this DQ8 competency.

DQ World Story

Honor
She believes your personal information and life are important enough to be honoured and protected.

The DQ Power of Honor treats each individual with great respect; as a valuable human being who deserves to have full control over the privacy in their digital lives. You will have the DQ Power of Honor when you truly understand the importance of privacy as a fundamental human right and when you have the set of knowledge, skills, attitudes and values described in the next section to handle all personal information with discretion, so as to protect one's own privacy as well as that of one's contacts.

Definition and Structure

The definition of DQ7: Privacy Management is the ability to handle with discretion all personal information shared online, to protect one's own and others' privacy.[2]

Structure	Description
Knowledge	Individuals understand privacy to be a human right; what personal information is and how it can be used, stored, processed and shared in digital platforms, along with strategies and tools that help them keep their personal information private and secure.
Skills	Individuals are able to develop behavioural and technical strategies to limit privacy violations, and are able to make good decisions around creating and sharing information and content of their own, as well as that of others.
Attitudes and Values	Individuals show respect for their own and other's privacy and personal information, treating these as valuable and as personal assets worth protecting.

8

DQ Index

Numbers are powerful. They tell us who we are and where we are. But numbers are also prophetic. They can dictate the direction of the future. Based on the *DQ Framework*, in this chapter, I will discuss how we can measure the level of DQ Digital Citizenship and Child Online Safety across individuals, schools/organizations, nations, and global, and more importantly, why we need to measure them. The goal of measurement must not be to rank them, but to empower them. Through the measurement, we can constantly be aware of where we are standing, and be checking which future heading toward to. So that it can guide us to readjust and redirect our direction for future to be better.

My vision for DQ score is to serve as an invisible hand to draw the line which is the bottom line. No single child will fall below that line. And above that line, every child can walk safely, run confidently, and fly boldly in this AI age.

Individuals: Are You Ready to Own a Smartphone?

I promised my son that he would get his first mobile phone if he got a minimum 115 on all eight DQ profile scores.

As a statistician, the first question I had when I started InfollutionZERO in 2010 was how to measure the levels of children's online safety and digital citizenship. It was not because I wanted to have another set of scores that children needed to be tested for, but because I wanted to develop a measure that could help both children and parents understand their level of cyber risks, strengths and weaknesses in digital citizenship. So that they can effectively empower children.

Wherever I go, the most frequently asked question for me, from parents is 'what is the age at which I should give a smartphone to my children?' I answered that they can when they get the minimum of 100 on DQ scores. It is like a driver's license. If you don't pass the driver's license test, you should not go behind the wheel, no matter how old you are. It is especially important for children who start to actively use smartphones and digital media, as this DQ score can indicate if they are ready to responsibly use these digital tools.

This year, my son turned thirteen years old. I promised that he would get his first mobile phone if he got a minimum 115 on all eight DQ profile scores. And he finally got his first mobile phone at his thirteenth birthday. Good DQ scores don't mean that he will be perfectly free from all cyber risks but at least they show that he has awareness of the various risks that he will encounter in the digital world and a basic knowledge, skills, and attitudes of how to deal with them and how to safely and responsibly use his devices and media as a good digital citizen.

Just like we can track our fitness progress by measuring body weight, BMI (body mass index) and other related metrics, I thought it could serve as a measure for us to visualize our current status of digital citizenship and online safety, and to reach the final goal of being a good digital citizen.

Along these lines, my design for the DQ Score was to help individuals achieve the following goals:

1. Plan the digital life you would like to have. Rather than blindly using technology and media as a slave, you can take control of your digital life.
2. Specify your goals for using digital technology, identify the things to avoid and awaken your conscientiousness to include your digital habits, to develop an orderliness in your digital lives.
3. Set a goal, make a schedule and develop micro-routines to encourage healthy micro-habits.
4. Incentivise yourself to improve your digital well-being and competency by allowing yourself to feel positive emotion when you see yourself moving towards higher DQ levels.
5. Build a positive relationship with your family and others.

So I call the DQ score a measure of an individual's digital readiness. In fact, the DQ score can be used not just for children, but for all age groups. Many of us as adults are also not qualified to have the mobile phone—those of us who have little understanding of DQ digital citizenship.

What is your DQ?

DQ Profile Scores

The DQ Profile scores provide a comprehensive summary of both the strengths and weaknesses of individuals' DQ digital citizenship, against a global standard. They measure the eight attributes of DQ digital citizenship competencies: (DQ1) digital citizen identity, (DQ2) balanced use of technology, (DQ3) cyberbullying management, (DQ4) personal cyber security management, (DQ5) digital empathy, (DQ6) digital footprint management, (DQ7) critical thinking, and (DQ8) privacy management. Each score is standardized; the global average is about 100 and a standard deviation of fifteen. The composite

DQ score is the average value of the eight DQ profile scores, which can be interpreted as a measure of an individual's digital readiness.

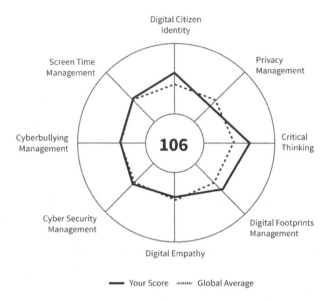

If your DQ score is:

- Below 85: You can potentially be exposed to one or more cyber risks or unhealthy habits of digital usage. It is recommended that you check your exposure to cyber risks.
- Between 85 and 100: You are at less than average compared to global standards in terms of corresponding DQ competency.
- Between 100 and 115: You are above average compared to your peers in terms of corresponding DQ competency. However, it is recommended to look at the overall profile with respect to all of their eight sub scores and understand your weaknesses and strengths. In case of children, you are advised to help the child to work on their weak areas, as we want children to have a holistic understanding of digital citizenship.
- Above 115: You can be considered to be relatively competent users in terms of corresponding DQ competency.

Schools/Organizations: Is Your School or Organization Digitally Resilient?

A robust digital citizenship education must include opportunities for assessment and feedback.

While working with schools in Singapore, I found that the biggest problem of the current digital citizenship education was the lack of a system to monitor the efficacy of education and cyber risk-control. A robust digital citizenship education must include opportunities for assessment and feedback. The assessment tools need to be comprehensive as well as adaptive to evaluate not only hard but also soft skills. Ultimately, the assessment should serve as a means of providing formative feedback that gives each individual child a better understanding of their own strengths and weaknesses, so that they may find their own paths to success. But it should also help school teachers and leaders design their digital citizenship education more effectively and proactively intervene and protect children at vulnerable to cyber risks.

So I designed the DQ School Report and Individual Scorecards as a starting point for conversation with children about their digital lives with teachers and parents. Parents, teachers and communities, all share responsibility in shaping children's digital habits and skills, and with these results, such conversations can be more honest and constructive. It is important to be positive and supportive, in order to gain valuable insights into our children's experiences that may otherwise never be shared. This way, we can offer concrete support for improving youth digital citizenship.

DQ Reports

The DQ Report provides a comprehensive summary of a child's digital life, including digital competency, usage, exposure to cyber dangers, personal strengths and digital support environment, as compared to other children within their nation. In addition, it also provides some practical recommendations to improve a child's DQ score, based on their current profile. They address the following questions:

- What are a child's DQ strengths and weaknesses, and how can teachers and parents encourage improvement?
- How balanced is a child's use of digital media and technology, and what proactive steps can teachers and parents take to improve this?
- Is a child at elevated exposure to cyber risks? How can teachers and parents proactively protect and intervene?

DQ Reports: Areas of Assessment

Area	Category
DQ Score	This category assesses the children's mastery of the eight key areas of Digital Citizenship.
Personal Strengths	This category assesses children's personal strengths, across the areas of global citizenship, social relationships, self-efficacy, self-regulation, emotional regulation, and balance of offline and online realities.
Balanced Use of Technology and Media	This category shows appropriately and in what ways children use digital devices and media; it provides useful information regarding their weekly screen time for entertainment use, device accessibility, digital media activity and social media usage.
Exposure to Cyber Risks	This category indicates children's exposure to cyber risks, including strangers online, game addiction, cyber bullying, cyber victimization, online sexual behaviour, and exposure to violent content.
Guidance and Support	This category reflects the degree of guidance children feel that they receive, with regard to proactive parental mediation, school computers and cyber safety education.

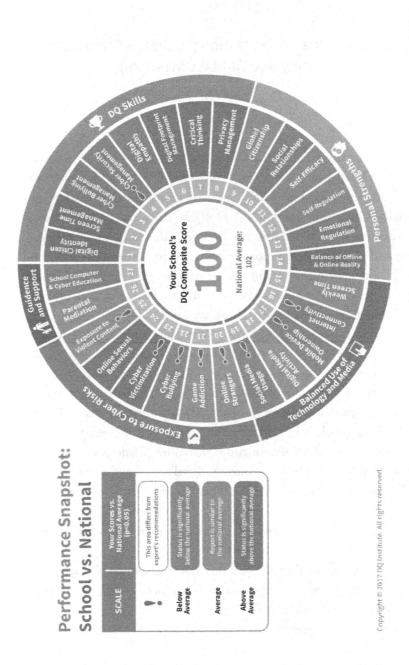

Performance Snapshot: School vs. National

SCALE	Your Scores vs. National Average (p<0.05)
!	This area differs from expert's recommendations
Below Average	Status is significantly below the national average
Average	Report is similar to the national average
Above Average	Status is significantly above the national average

DQ Skills
- Digital Empathy
- Digital Footprint Management
- Critical Thinking
- Privacy Management

Personal Strengths
- Global Citizenship
- Social Relationships
- Self-Efficacy
- Self-Regulation
- Emotional Regulation

Balanced Use of Technology and Media
- Balance of Offline & Online Reality
- Weekly Screen Time
- Internet Connectivity
- Mobile Device Ownership
- Digital Media Activity
- Social Media Usage

Exposure to Cyber Risks
- Online Strangers
- Game Addiction
- Cyber Bullying
- Cyber Victimization
- Online Sexual Behaviors
- Exposure to Violent Content

Guidance and Support
- Parental Mediation
- School Computer & Cyber Education
- Digital Citizen Identity
- Screen Time Management
- Cyber Bullying
- Cyber Security Management

Your School's DQ Composite Score
100
National Average: 102

Figure x. Sample of the DQ School Report: Performance snapshot of various dimensions of digital citizenship, exposure to cyber risks and others.

Nations: What is the National Level of Child Online Safety and Digital Citizenship?

Every nation's basic digital readiness must start from ensuring the safety of children online and digital citizenship education

On 6 February 2018, which was the Safer Internet Day in that year, we first published the *2018 DQ Impact Report* in association with the World Economic Forum.[1] This report examined online safety and digital citizenship among 38,000 eight- to twelve-year-olds across twenty-nine countries. Our key finding at that time was that 56 per cent of children were exposed to at least one cyber risk (including risks like cyberbullying and online grooming). This report created an impetus to expand and deepen our knowledge of the risks that children face, as well as what resources can act as protective factors.

From the experience of #DQEveryChild since 2017, I realized that educating children, alone, can't be enough. It is critically important to build healthy digital environments around children that connect every stakeholder, including schools, families, communities, ICT companies and governments. My old-time genius colleague, Masali Hancock, who was a founding member of iKeepSafe and is now the CEO of EP3Foundation, shared her wisdom with me that child safety must include ensuring digital safety and security across all four points of digital contact for children:

1. where they connect, i.e., children's environment (such as school or family or community),
2. the device they connect through (e.g. mobile, PCs),
3. the Internet connection they connect with (e.g. Internet Service Provider),
4. the software/apps they use (such as social media, game apps, etc.).

When any of these stakeholders fall short on maintaining the DQ standard, children can be easily fall into cyber-dangers and the digital education ecosystem can be ruined.

I always believe that every nation's basic digital readiness must start from ensuring the safety of children online and digital citizenship education. To support nations, we need a global index on child online safety and digital citizenship, to build the global standards for all stakeholders around children's digital environment in each nation, starting from children and extending to their immediate supporters, such as parents and teachers, and further, to community, ICT companies and the government.

With this understanding, we have developed the new national level index in 2020, the *Child Online Safety Index* (COSI).[2] The COSI was the world's first real-time analytic platform to help nations better monitor the status of their children's online safety and digital citizenship.

The COSI is measured across six pillars which form the COSI framework. Pillars one and two, Cybers Risks and Disciplined Digital Use, relate to wise use of digital technology. Pillars three and four, Digital Competency and Teaching & Guidance are related to empowerment. The final two pillars relate to Infrastructure; these are the pillars of Social Infrastructure and Connectivity.

In other words, the COSI measures if the current digital ecosystem in a particular nation support every child to have an equal opportunity to thrive in the digital future safely, securely, and well. Every child needs to be:

- able to safely, responsibly and ethically use digital devices and media with digital citizenship skills (Cyber risks, Disciplined Digital Use, and Digital Competency)
- supported by parents, caregivers and educators (Teaching & Guidance)

- protected against various cyber risks through government policies, civic engagement, and ethical business practices (Social Infrastructure)
- able to fully access and use digital devices and media (Connectivity)

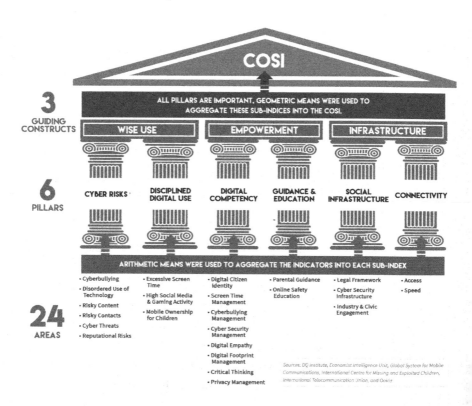

Using the index, we hope that nations will be able to identify areas of improvement for their children's online safety issues in terms of these six pillars. Global benchmarking will make targeting those areas more effective and improve deployment of relevant programmes and initiatives. Actors can then effectively coordinate efforts to enhance child online safety and digital citizenship within their nations, with the ability to measure national progress.

Moreover, I wanted this COSI to always be live and participatory, not static statistics that are only measured annually. So we linked the COSI with DQ assessment tools and its global database; the COSI will now be automatically updated as nations progress with their initiatives to improve child online safety. I envisioned this COSI to trigger every nation to implement DQ education and assessment, and to level-up the DQ of all stakeholders in entire digital education ecosystem. I believed in this method; even though DQ starts with children, it will lead to building more inclusive, creative and responsible smart nations that can thrive in the future.

COSI REFLECTS #DQEVERYCHILD PROGRESS IN REAL TIME

Global: Shouldn't We Include Well-Being and Sustainability in Measurement for Digital Economy and Competitiveness?

Child online safety is a serious barometer of the sustainability and trust dimension of the digital economy.

In January 2020, I got an unexpected email from Princess Nouf Muhammad Al Saud of Saudi Arabia. She asked if I can serve as the international lead for 'Digital Economy for G20 Civil Society (C20)'[3] under Saudi Arabia's presidency, the host for the G20 2020. What? Digital Economy? For G20 leaders?

Following this, I got the amazing opportunity to work with the G20 Digital Economy Task Forces and more than 750 civic societies globally, to discuss various policy issues related to digital economy. Most importantly, in June 2020, I was given an opportunity to present on behalf of the global civil society, at the G20 Digital Economy Task Force meeting, that was aimed at developing the common framework to measure the success of digital economy.

I felt that it was one of my last tasks in my ten-year social impact journey for child online protection, as well as my three-year #DQEveryChild journey; just as I had promised to the World Economic Forum and its 100 other global partners—to bring digital citizenship education to children around the world from 2017 to 2020. I shared my initial three ten-year goals that I had made for myself, in 2010: 1) to develop a global standards framework that every nation can adopt, 2) a global child education programme that every child can use and 3) a global index that every nation shall pay attention to.

I published the *DQ Framework* in 2015; I developed #DQEveryChild in 2017, and I published the *Child Online Safety Index* in 2020. I wanted to make COSI into a global index that every nation would pay attention to. So the June G20 Digital Economy presentation was like a meant-to-be, so I could accomplish my ten-year goals, as well as my promise with #DQEveryChild partners.

The year 2020, with the COVID-19 pandemic, was an unprecedented time in the course of human history. The COVID-19 has hard-pushed almost everyone in global society, including governments, businesses and individuals, to fast-track our entry into the Digital Economy. 'Go Digital' was no longer an agenda of government and ICT industry. While the world scrambles for effective solutions to the COVID-19 pandemic, all of us have become increasingly reliant on digital technologies to connect each other, as face-to-face contact is minimized with nations' lockdowns and social distancing measures being enforced. Indeed, almost all nations have been rushing into digital transformation at a super-fast pace.

Going through the COVID-19 situation, I seriously began to worry about who will bravely raise the concerns around such rushed digitization. Usually, the 'negative what-ifs' discussion about potential negative consequences of technology—Infollution—has not been welcomed by government and industry leaders, as the global and national agendas are focused around 'digital transformation'. We expect widespread Infollution that includes children's exposure to cyber risks, digital surveillance, increasing job loss due to automation, privacy and data ownership issues, cyber-security and safety, monetization schemes based on cognitive deceit, and many others. In this time of the COVID-19 crisis, life or death and being connected or unconnected to the society make these other concerns seem irrelevant or luxurious, making them secondary priorities.

So I decided to broaden the COSI agenda to be included into the well-being, ethics, and sustainability agenda of the digital society, when I was preparing for my speech at the G20 Digital Economy Task Force meeting to develop the Measurement for Digital Economy.

To me, this speech was a closure of the first chapter of my social impact journey. I will close this book with my speech at that meeting. I am starting a completely new second chapter of my social impact journey with partners who will co-create the DQ Global Standards for our children. If you believe in the power of our children, join us.

Systemic Collaboration with C20 on the G20 Measurement of the Digital Economy

Roadmap Toward a G20 Common Framework for Measuring the Digital Economy

Yuhyun Park, PhD
International Lead, C20 Digital Economy Working Group
Founder of DQ Institute

Hello everyone,

I'm Yuhyun Park, the founder of DQ Institute, serving as the international lead for the C20 digital economy working group. It is privilege to speak on this topic of measurement of digital economy on behalf the global civil society.

I am a statistician and expert, who has been working on assessment and indexing of digital skills and digital economy from academic and civic sectors. Numbers are powerful. They tell us who we are and where we are. But numbers are also prophetic. They can dictate the direction of the future. So what we discuss today, can define the future that you and I, your children and mine will live in.

At this opportunity, I would like to suggest a few ways in which global civil society can support and collaborate with G20 effort of measurement of digital economy in a systematic way, across the five elements of the 'roadmap'—from definition and indicators to dissemination and implementation.

1. Common Definition and Dissemination

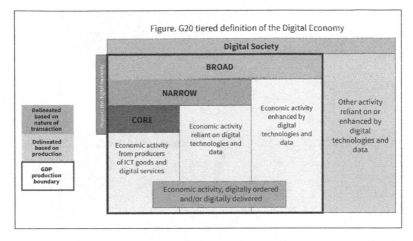

Figure. G20 tiered definition of the Digital Economy

When it comes to the definition of the digital economy, we want to again highlight that the Digital Economy is about 'people', not about 'technology'; the digital economy is intrinsically rooted in individual lives, at the core of its working mechanisms. For instance, Alexa listens to your family conversations, YouTube educates your children, instead of school teachers, and Siri views your confidential work e-mails—such personal data becomes the fuel which drives AI R&D. We often hear, 'data is the new oil', which means that personal data has become the most important commodity in the Digital Economy.

So we believe that individual and societal empowerment and well-being must be considered while defining 'digital economy'. For these reasons, C20 welcomes the new concept of a 'digital society'. It is a great evolution from the previous, narrower definition, restricted to economic activities related to technology. C20 will work with G20 in dissemination of this new definition, and support to strengthen its measurement to be aligned with UN 2030 sustainable development goals.

2. Indicators for 'Human-centered' Digital Society

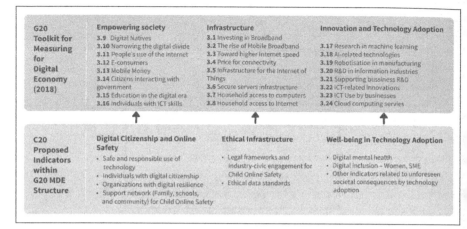

Along those lines, we have made the following suggestions; of the indicators to be included in the measurement of digital economy. Our suggested indicators are based on the structure of the 2018 G20 measurement Toolkit,[4] which is beautifully done. Moreover, these indicators further support the new definition of 'digital society' by enhancing individual and societal well-being, empowerment and resilience, that have been discussed throughout today's meeting, and they are aligned with the 2020 G20 Saudi presidency's priority of 'human-centred' and 'trustworthy' technology.

The first pillar is 'empowering society'. C20 believes it is important to strengthen it with the indicators related to digital citizenship and online safety of individuals and organizations, starting from children, women and small and mid-size enterprises or SMEs.

The second pillar is 'infrastructure', which currently covers the hard infrastructure and connectivity-related issues, and can be enhanced by including soft infrastructure. Specifically, we suggest that the indicators related to 'ethical infrastructure' also be taken into consideration and used for guidance, which includes information on child online safety, personal data protection, and ethical data standards.

The third pillar is 'innovation and technology adoption'. We believe that it can also be strengthened by the indicators related to well-being in technology adoption, which covers digital inclusion, digital mental health, and other human rights aspects.

There are already tested and proven definitions for these indicators, measurement tools and methodology, which can concretely contribute to the G20 effort for measurement in C20, academia, and other engagement groups.

This is an example of how C20 can systematically work on the development of indicators.

In terms of the skills indicator, the current issue is that there is no globally shared understanding of what terms like 'digital skills' or 'ICT skills' mean. The current indicators for ICT skills and related measurement instruments are often narrowly defined as basic operational digital skills on how to use software and devices.

But digital transformation is creating demand for a comprehensive set of new digital skills, necessary for the digital economy. And there are existing, agreed, tested and proven global standards frameworks, common language for digital skills such as digital intelligence framework that was used and endorsed by IEEE Standards Association, OECD, and the World Economic Forum. These capacities include not only technical skills, but also cognitive, and socio-emotional skills related to digital economy, such as online privacy and the spread of digital misinformation.

So there are common measurements, indicators and instruments that are readily available to deploy in C20 and other engagement groups.

Please visualize the digital society. Are we happy and empowered? Are we safe and secure?

Yes? But the evidence and research tell otherwise.

3. Example: Indicators

Jobs and Growth

3.25 Jobs in the Information Industries
3.26 Jobs in ICT occupations
3.27 ICT workers by gender
3.28 E-Commerce
3.29 Value added in information industries
3.30 The extended ICT footprint
3.32 ICT and productivity growth
3.33 ICT and global value chains
3.34 Trade and ICT Jobs
3.35 ICT goods as a percentage of merchandise trade
3.36 Telecommunications, computer, and information services as a percentage of services trade

Section	Indicator name	Underpinning data source
Jobs	2.1.1 Jobs in digital intensive sectors and information industries	Labour Force Surveys
	2.2.1 Jobs in ICT task-intensive and ICT specialist occupations	Labour Force Surveys
	2.2.2 ICT professionals adn technicians by gender	Labour Force Surveys
Skills	3.1.1 Selected ICT skills by gender	Surveys of OCT usage by individuals / of modules in LFS
	3.2.1 ICT task intensity of jobs, by gender	OECD Survey of Adult Skills (PIAAC)
	3.2.1 ICT usage in school	OECD PISA
	3.3.2 Students' reported ICT capabilities, by gender	OECD PISA
	3.4.1 Tertiary graduates in natural sciences, engineering, ICTs, and content fields of education	Labour Force Surveys
	3.4.2 Tertiary graduates in NSE & ICT, by gender	Labour Force Surveys
Growth	4.1.1 Value added by information industries	National Accounts
	4.1.3 Value added by digitally intensive sectors	National Accounts
	4.2.1 ICT investment by asset	National Accounts
	4.3.1 ICT goods exports and imports	Merchandise trade data
	4.2.1 Digitally-deliverabale services exports and imports	Trade in Services data

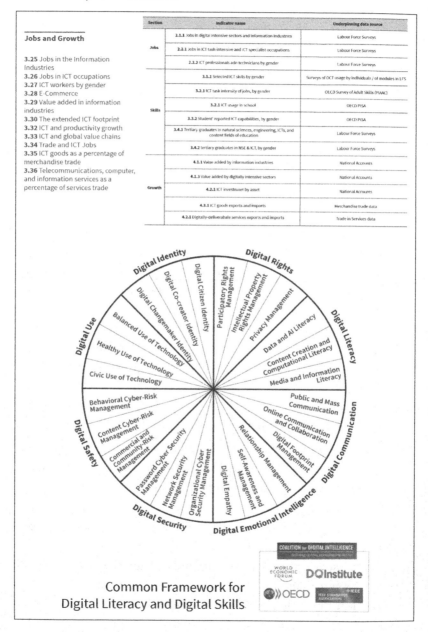

Common Framework for
Digital Literacy and Digital Skills

One of the most important topics that C20 discussed this year, was children's high exposure to cyber risks. It was shown by a global study across thirty countries, that 60 per cent of children aged eight to twelve are exposed to at least one cyber risk—that includes cyberbullying, gaming addiction, fake news, cyberattacks and others.

This is one of key areas that directly threaten the sustainability of the digital economy. Citizen's trust toward government, ICT and media industries have been decreasing every year for the last few years. Even Apple board members call mobiles and children paired topics. Child online safety is a serious barometer of the sustainability and trust dimension of the digital economy.

We emphasize that the G20 measurement should capture this social impact of unintended consequences of technology in the digital society, so that we can monitor and improve our digital economy, holistically. This is one example of why we strongly suggest that indicators for digital citizenship and online safety are included in the G20 effort for measurement of digital economy, ideally within this year, or going forward.

Such an inclusion of better indicators won't take time if we work together. We suggest G20 to collaborate with C20 and other engagement groups by incorporating their best practices. Related to *Child Online Safety Index*, DQ Institute has the largest global data bank and research-based methodology, related to online safety and digital citizenship across eighty countries; we are exploring and discussing with various organizations including UN, OECD, ITU, IEEE, UNICEF and others, about building a global databank to develop common measures for child online safety and digital citizenship.

All of these best practices and national-level actions can enhance the comprehensiveness of the G20 measurement.

We can further support this effort by working with major international coalitions around the world. It will immediately disseminate to grassroot players around the world and make these indicators a real measure.

4. Dissemination of Indicators and Institutional arrangement and capabilities

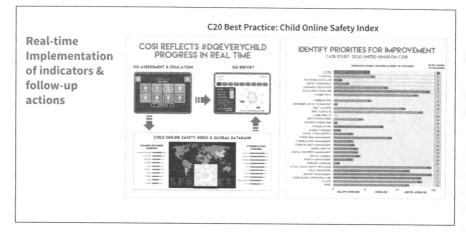

It is not just international organizations like ITU, UN, or World Economic Forum; there are many similar global coalitions that include civil societies, academia and companies in their network, which need the impact and success measures.

We can go one step further. Imagine if the G20 Measurement for Digital Economy could be monitored, measured and tracked in real-time through collaborative implementations of various indices, and real-life business, and activities of individuals and organizations?

DQ Institute ran #DQEveryChild for teaching children about digital skills with real-time assessment, together with 100 international partners. Children's data and analytics go directly to their parents and teachers, and they are also reflected in the *Child Online Safety Index*—the global report as well as the national report—on the left side, in real-time.

Visualize if G20 measurements can be also included in this report aimed at players across the sectors. Such real-time measurement and data visualization can impact every layer in society—individuals, organizations, governments and global societies—in a powerful way. These numbers will not sit on the policy makers' table, but can change people's lives in real-time. Truly digital.

Current G20, OECD and national statistical offices' collaborative approach in developing this measurement is great. But we can be more than great. We can be *fantastic*. I would like to suggest that we should have a continuing conversation and build a broader coalition that engages C20 and other engagement groups in the development and deployment of the G20 effort for measurement of digital economy, as the next step.

I will return the mic to the chair.

Thank you for listening.

Thank you

There are so many people that I want to express my deep gratitude.

I always remember the people who helped me in difficult times. Dae-Hyeuk Park, Jung-Hoon Kim, Dong-Ho Kim were my first supporters when I started infollutionZERO. Dong-Joon Cho, Josh Jackman and Yumi Kim who joined me from the day 1 and have been working with me until today. My sister, Miah Park, and her friends volunteered their time and efforts, which enabled the start of the iZ HERO project. Dr Lee Sang-Hee, Yoon Jong-Rok, So Hoon Lim, Kwok Mean Luck, Mei Lin Fung, Hwang Wu-Yeo, Yong-Rok Yoon, Su Fen Toh, YM Muntean, and so many people appeared to this journey like God sent angels and have been supporting the cause of #DQEveryChild with full trust. My DQ team, partners, and supporters are the real heroes. Despite of my terrible leadership, they have been patient enough to endure me and they have been walking with me.

Dr Un-Tae Park, my dad, taught me to choose a life path that I believe without fear. The Ahn-Sei Foundation that he founded in 1985 provided a foundation for DQ Institute. Dr Pack-Jae Cho, my father-in-law, who first coined the term—Infollution have inspired and supported me. Hae-Soon Suh, my mom, and Dr Kye-Soon Im, mother-in-law empowered me to be who I am as a mom, as a daughter, and as a career woman. Isaac Cho—my son and Kate Cho—my daughter are my heart and soul. Dr Nam-Joon Cho, my husband, my best friend, and my love. Thank you. Lastly, all glory belongs to my savior, Jesus Christ.

References

Reflection

1. Herskovitz, J., & Kim, C. (9 November 2009). 'South Korea Seeks New Laws After Brutal Rape of Child'. *Reuters*. https://www.reuters.com/article/idININdia-43802720091109
2. Satell, G. (4 September 2014). 'A Look Back At Why Blockbuster Really Failed And Why It Didn't Have To'. *Forbes*. https://www.forbes.com/sites/gregsatell/2014/09/05/a-look-back-at-why-blockbuster-really-failed-and-why-it-didnt-have-to/?sh=dc16bdf1d64a
3. Tedeneke, A. (26 September 2018). OECD, IEEE and DQI Announce Platform for Coordinating Digital Intelligence Across Technology and Education Sectors. World Economic Forum. https://www.weforum.org/press/2018/09/oecd-ieee-and-dqi-announce-platform-for-coordinating-digital-intelligence-across-technology-and-education-sectors/
4. DQ Institute (2020). *#DQEveryChild*. https://www.dqinstitute.org/dqeverychild/
5. DQ Institute (2020). *DQ World*. https://www.dqworld.net/
6. C20 Saudi Arabia (2020). *2020 Civil Society 20 Policy Pack*. 11 2020 https://civil-20.org/2020/wp-content/uploads/2020/06/2020-C20-Policy-Pack.pdf
7. McCabe, K. (10 October 2020). *New Standard Will Help Nations Accelerate Digital Literacy and Digital Skills Building*. IEEE Standards Association. https://beyondstandards.ieee.org/working-groups/new-standard-will-help-nations-accelerate-digital-literacy-and-digital-skills-building/

8. DQ Institute (10 October 2019). *Welcome to DQ Day 2019.* https://www.dqinstitute.org/dq-events/
9. Newell, M. (2012). *Great Expectations.* BBC Films.

Chapter 1

1. Schwab, K. (2016). *The Fourth Industrial Revolution.* World Economic Forum.
2. Harari, Y. N. (2014). *Sapiens: A Brief History of Humankind.* Random House.
3. Harari, Y. N. (2016). *Homo Deus: A Brief History of Tomorrow.* Random House.
4. Hutchison III, C. A., Chuang, R., Noskov, et al. (2016). 'Design and synthesis of a minimal bacterial genome'. *Science*, 251(6280). https://science.sciencemag.org/content/351/6280/aad6253
5. Polkinghorne, J. (2007). *One World: The Interaction of Science and Theology.* Templeton Foundation Press.
6. Tedeneke, A. (26 September 2018). OECD, IEEE and DQI Announce Platform for Coordinating Digital Intelligence Across Technology and Education Sectors. World Economic Forum. https://www.weforum.org/press/2018/09/oecd-ieee-and-dqi-announce-platform-for-coordinating-digital-intelligence-across-technology-and-education-sectors/
7. Luntz, S. (16 March 2016). Microsoft's Chatbot Quickly Converted To Bigotry By The Internet. *ISL Science.* https://www.iflscience.com/technology/microsofts-chatbot-converted-bigotry/
8. United Nations. *Universal Declaration of Human Rights.* https://www.un.org/en/universal-declaration-human-rights/
9. Russell, B. (1950). *What Desires Are Politically Important?* Nobel Prize for Literature Acceptance Speech. Stockholm: Sweden.
10. Jensen, F. (1982). *C.G. Jung, Emma Jung, E., & Toni Wolff: A Collection of Remembrances.* The Analytical Psychology Club of San Francisco.
11. Jung, C.G.(1992). C.G. Jung Letters, Volume 1: 1906-1950 v. 1. Bollingen Series.
12. Lewis, C.S. (1940). *The Problem of Pain.* The Centenary Press.
13. Eisenhower Fellowships. https://www.efworld.org/

14. Nadella, S., Shaw, G., & Nichols, J.T. (2017). *Hit Refresh: The Quest to Rediscover Microsoft's Soul and Imagine a Better Future for Everyone.* HarperCollins

15. Markoff, J. (2015). *Machines of Loving Grace: The Quest for Common Ground Between Humans and Robots.* HarperCollins.

16. Forrester (2020). *Future of Work.* https://go.forrester.com/future-of-work/

Chapter 2

1. DQ Institute (2020). *Child Online Safety Index.* https://www.dqinstitute.org/child-online-safety-index/

2. DQ Institute (2018). *2018 DQ Impact Report.* https://www.dqinstitute.org/2018dq_impact_report

3. 7 News (2020). 'Coronavirus Restrictions: 11-Year-Old Boy Leaps From Building With Sister Trying To Recreate Video Game'. https://7news.com.au/lifestyle/health-wellbeing/coronavirus-restrictions-11-year-old-boy-leaps-from-building-with-sister-trying-to-recreate-video-game-c-1038197

4. Singer, N. (12 February 2018). 'Tech's Ethical 'Dark Side': Harvard, Stanford and Others Want to Address It'. *The New York Times.* https://www.nytimes.com/2018/02/12/business/computer-science-ethics-courses.html

5. UNICEF (1989). *United Nations Convention on the Rights of the Child (UNCRC).* https://www.unicef.org.uk/wp-content/uploads/2010/05/UNCRC_united_nations_convention_on_the_rights_of_the_child.pdf

6. DQ Institute (2020). *#DQeveryChild.* https://www.dqinstitute.org/dqeverychild/

7. Friedman, T. L. (2016). *Thank You for Being Late: An Optimist's Guide to Thriving in the Age of Accelerations.* Farrar, Straus and Giroux.

8. Munro, K. (2017, March 3). 'Don't teach your kids coding, teach them how to live online'. *Sydney Morning Herald.* https://www.smh.com.au/national/nsw/dont-teach-your-kids-coding-teach-them-how-to-live-online-20170324-gv5e9r.html.

9. OECD (2018). *OECD Future of Education and Skills 2030.* https://www.oecd.org/education/2030-project/

10. OECD (2018). *Megatrends influencing the future of education*. https://www.oecd.org/education/2030-project/teaching-and-learning/learning/megatrends/

Chapter 3

1. Marr, B. (2020). 'What Is GPT-3 And Why Is It Revolutionizing Artificial Intelligence?' *Forbes*. https://www.forbes.com/sites/bernardmarr/2020/10/05/what-is-gpt-3-and-why-is-it-revolutionizing-artificial-intelligence/?sh=5e77aa91481a

2. Robertson, M.R. (2017). 'Twenty years on from Deep Blue vs Kasparov: how a chess match started the big data revolution'. *The Conversation*. https://theconversation.com/twenty-years-on-from-deep-blue-vs-kasparov-how-a-chess-match-started-the-big-data-revolution-76882

3. DeepMind (2016). *The Google DeepMind Challenge Match*. https://deepmind.com/alphago-korea

4. *The Economist* (10 March 2005). 'The New Pharaohs'. https://www.economist.com/business/2005/03/10/the-new-pharaohs

5. Blodget, H. (19 January 2012). 'CEO of Apple Partner Foxconn: "Managing One Million Animals Gives Me A Headache"'. *Business Insider*. https://www.businessinsider.com/foxconn-animals-2012-1

6. Wallach, W. (2015). *Dangerous Master: How to Keep Technology from Slipping Beyond Our Control*. Basic Books.

7. Markoff, J. (2015). *Machines of Loving Grace: The Quest for Common Ground Between Humans and Robots*. Ecco Press.

8. World Health Organisation (2018). *Addictive behaviours: Gaming disorder*. https://www.who.int/news-room/q-a-detail/addictive-behaviours-gaming-disorder

9. DQ Institute (2020). *Child Online Safety Index*. https://www.dqinstitute.org/child-online-safety-index/

10. *The New York Times* (16 January 2021). *Children's Screen Time Has Soared in the Pandemic, Alarming Parents and Researchers*. https://www.nytimes.com/2021/01/16/health/covid-kids-tech-use.html

11. Nadella, S., Shaw, G., & Nichols, J.T. (2017). *Hit Refresh: The Quest to Rediscover Microsoft's Soul and Imagine a Better Future for Everyone*. HarperCollins

12. Kay, A. (2019). 'To what extent was it possible to build a digital computer during ancient Rome?' https://www.quora.com/To-what-extent-was-it-possible-to-build-a-digital-computer-during-ancient-Rome.
13. Orwell, G. (1949). *1984*. Secker & Warburg.

Chapter 4

1. Yad Vashem. 'Yad Vashem – The World Holocaust Remembrance Center'. https://www.yadvashem.org/
2. Arendt, H. (1963). *Eichmann in Jerusalem: A Report on the Banality of Evil*. Viking Press.
3. Brooker, K. (2018). 'I Was Devastated': Tim Berners-Lee, The Man Who Created The World Wide Wed, Has Some Regrets. *Vanity Fair*. https://www.vanityfair.com/news/2018/07/the-man-who-created-the-world-wide-web-has-some-regrets
4. The Beatles (1970). *I Me Mine (Song)*. Apple Music
5. Pavlov, I. P. (1927). *Conditioned Reflexes: An Investigation of the Physiological Activity of the Cerebral Cortex*. Oxford University Press
6. Vosoughi, S., Roy, D., & Aral, A. (2018). 'The Spread of True and False News Online'. *Science 359* (6380): 1146-1151.
7. Meyer, R. (2018, March 8). 'The Grim Conclusions of the Largest-Ever Study of Fake News'. *The Atlantic*. https://www.theatlantic.com/technology/archive/2018/03/largest-study-ever-fake-news-mit-twitter/555104/
8. Abba (1980) *Winner Takes it All (Song)*. Polar.
9. Meyer, R. (28 June 2014). 'Everything We Know About Facebook's Secret Mood Manipulation Experiment'. *The Atlantic*. https://www.theatlantic.com/technology/archive/2014/06/everything-we-know-about-facebooks-secret-mood-manipulation-experiment/373648/
10. Fussell, S. (12 October 2018). 'Alexa Wants to Know How You're Feeling Today'. *The Atlantic*. https://www.theatlantic.com/technology/archive/2018/10/alexa-emotion-detection-ai-surveillance/572884/
11. Popkin, H.A.S. (13 January 2010). 'Privacy is dead on Facebook. Get over it'. *NBC News*. https://www.nbcnews.com/id/wbna34825225

12. BBC (10 January 2019). 'Cambridge Analytica parent firm SCL Elections fined over data refusal'. https://www.bbc.co.uk/news/technology-46822439

13. Cadwalladr, C. & Graham-Harison, E. (17 March 2018). 'Revealed: 50 million Facebook profiles harvested for Cambridge Analytica in major data breach'. *The Guardian.* https://www.theguardian.com/news/2018/mar/17/cambridge-analytica-facebook-influence-us-election

14. Dreyfus, E. (24 July 2019). 'Netflix's The Great Hack Brings Our Data Nightmare to Life'. *Wired.* https://www.wired.com/story/the-great-hack-documentary/

15. Amer, K., & Noujaim, J. (2019). *The Great Hack.* Netflix

16. Lapowsky, I. (25 January 2019). 'One Man's Obsessive Fight to Reclaim His Cambridge Analytica Data'. *Wired,* https://www.wired.com/story/one-mans-obsessive-fight-to-reclaim-his-cambridge-analytica-data/

17. *European Commission.* 'What does the General Data Protection Regulation (GDPR) govern?' https://ec.europa.eu/info/law/law-topic/data-protection/reform/what-does-general-data-protection-regulation-gdpr-govern_en

18. Lapowsky, I. (25 January 2019). 'One Man's Obsessive Fight to Reclaim His Cambridge Analytica Data'. *Wired,* https://www.wired.com/story/one-mans-obsessive-fight-to-reclaim-his-cambridge-analytica-data/

19. *Datareportal* (2020). *Digital Around the World.* https://datareportal.com/global-digital-overview#:~:text=Roughly   per   cent204.66 per cent20billion per cent20people per cent20around,twelve per cent20months per cent20to per cent20October per cent202020.

20. Naver (15 April 2011). 뉴스캐스트 시민단체 모니터링단 출범식 https://m.blog.naver.com/naver_diary/150106726083

21. 아시아경제 (29 November 2018). '공정성 논란' 네이버 뉴스, 사람 개입 없이 적절히 운영. https://www.asiae.co.kr/article/2018112910565069885

22. BBC (2016) *EU Referendum.* https://www.bbc.co.uk/news/politics/eu_referendum

23. Stone, Z. (10 October 2016). '11 Times Mark Zuckerberg Kept It Real'. *Forbes.* https://www.forbes.com/sites/zarastone/2016/10/10/11-times-mark-zuckerberg-kept-it-real/?sh=697626d735d4

24. *The New York Times* (2 August 1964). 'The Churchill Spirit – In His Own Words'. https://www.nytimes.com/1964/08/02/archives/the-churchill-spiritin-his-own-words.html

25. Gorlick, A. (2009). 'Media multitaskers pay mental price, Stanford study shows'. *Stanford News*. https://news.stanford.edu/news/2009/august24/multitask-research-study-082409.html

26. Kuznek, J.H. & TItsworth, S. (2013). 'The Impact of Mobile Phone Usage on Student Learning'. *Communication Education*, 62(3), 233-252.

27. Bates, S (25 October 2018). 'A decade of data reveals that heavy multitaskers have reduced memory, Stanford psychologist says'. *Stanford News*. https://news.stanford.edu/2018/10/25/decade-data-reveals-heavy-multitaskers-reduced-memory-psychologist-says/

28. *Quora*. 'What did Einstein mean by stating, "The monotony & solitude of quiet life stimulates the creative mind?"' https://www.quora.com/What-did-Einstein-mean-by-stating-The-monotony-solitude-of-quiet-life-stimulates-the-creative-mind

29. Twenge, J.M. (9 September 2017). 'Has the Smartphone Destroyed a Generation'. *The Atlantic*. https://www.theatlantic.com/magazine/archive/2017/09/has-the-smartphone-destroyed-a-generation/534198/

30. Bamwita, T. (2014, May 13). 'On war against global terrorism and our ideologies'. *The New Times*. https://www.newtimes.co.rw/section/read/75268

31. Arendt, H. (1963). *Eichmann in Jerusalem: A Report on the Banality of Evil*. Viking Press.

32. Williams, J. (2018, May 27). 'Technology is driving us to distraction'. *The Guardian*. https://www.theguardian.com/commentisfree/2018/may/27/world-distraction-demands-new-focus)

Chapter 5

1. Brynjolfsson, E., & McAfee, A. (2014). *Race Against The Machine: How the Digital Revolution is Accelerating Innovation, Driving Productivity, and Irreversibly Transforming Employment and the Economy*, Digital Frontier Press. https://www.google.co.in/books/edition/Race_Against_the_Machine/6O-MBAAAQBAJ?hl=en&gbpv=1

2. OECD. *Future of Education and Skills 2030*. https://www.oecd.org/
    education/2030-project/
3. Tedeneke, A. (26 September 2018). 'OECD, IEEE and DQI
    Announce Platform for Coordinating Digital Intelligence Across
    Technology and Education Sectors'. World Economic Forum. https://
    www.weforum.org/press/2018/09/oecd-ieee-and-dqi-announce-
    platform-for-coordinating-digital-intelligence-across-technology-and-
    education-sectors/
4. United Nations. *The 17 Goals*. https://sdgs.un.org/goals
5. United Nations. *Universal Declaration of Human Rights*. https://www.
    un.org/en/universal-declaration-human-rights/
6. OECD. *Measuring Well-being and Progress: Well-being Research* https://
    www.oecd.org/statistics/measuring-well-being-and-progress.htm.
7. Park, Y. (3 August 2016). 'The Fourth Industrial Revolution Awakens
    the Importance of the Human Spirit'. *Huffington* Post. https://www.
    huffpost.com/entry/the-fourth-industrial-rev_b_11325636
8. Park, Y. (2016). *The Future of Human Intelligence*. https://www.
    youtube.com/watch?v=I6qR7hx89VU
9. Estlin, P. (2019). *R U DQ?* https://www.gresham.ac.uk/lectures-and-
    events/digital-skills-crisis-opportunity
10. DQ Institute (2019). *DQ Global Standards Report 2019*.
    https://www.dqinstitute.org/wp-content/uploads/2019/03/
    DQGlobalStandardsReport2019.pdf
11. Treschow, M. (1994). 'The Prologue to Alfred's Law Code: Instruction
    in the Spirit of Mercy'. *Florilegium (13)*. https://journals.lib.unb.ca/
    index.php/flor/article/download/18463/20304/24489
12. Pawson, D. (2019). *The Makers Instructions: A new look at the 10
    Commandments.*   https://www.davidpawson.org/books/the-makers-
    instructions/
13. DQ Institute (2019). *Global Launch of Digital Intelligence Day on
    October 10th Co-Creating Global Standards for Digital Literacy, Skills,
    and Readiness.*   https://www.dqinstitute.org/news-post/global-launch-
    of-digital-intelligence-day-on-october-10th-co-creating-global-
    standards-for-digital-literacy-skills-and-readiness/
14. Goleman, D. (1999). *Working with Emotional Intelligence*. Bloomsbury
    Publishing PLC.

15. Goleman, (9 December 2019). 'The 8 pieces of digital DNA we need to thrive in the AI age'. World Economic Forum. https://www.weforum. org/agenda/2019/12/digital-intelligence-artificial-intelligence-ethics/
16. Park, Y. (6 September 2016). '8 digital life skills all children need – and a plan for teaching them'. World Economic Forum. https://www. weforum.org/agenda/2016/09/8-digital-life-skills-all-children-need-and-a-plan-for-teaching-them/
17. Twenge, J.M. (9 September 2017). 'Has the Smartphone Destroyed a Generation'. *The Atlantic.* https://www.theatlantic.com/magazine/archive/2017/09/has-the-smartphone-destroyed-a-generation/534198/
18. Park, Y. (6 September 2016). '8 digital life skills all children need – and a plan for teaching them'. World Economic Forum. https://www. weforum.org/agenda/2016/09/8-digital-life-skills-all-children-need-and-a-plan-for-teaching-them/
19. Tedeneke, A. (26 September 2018). 'OECD, IEEE and DQI Announce Platform for Coordinating Digital Intelligence Across Technology and Education Sectors'. World Economic Forum. https:// www.weforum.org/press/2018/09/oecd-ieee-and-dqi-announce-platform-for-coordinating-digital-intelligence-across-technology-and-education-sectors/
20. Future.Now. *Empowering everyone to thrive in a digital UK.* https:// futuredotnow.uk/

Chapter 6

1. DQ Institute (2020). *Child Online Safety Index.* https://www. dqinstitute.org/child-online-safety-index/
2. Mielach, D. (2012, April 19). 'We Can't Solve Problems by Using the Same Kind of Thinking We Used When We Created Them'. *Business Insider.* https://www.businessinsider.com/we-cant-solve-problems-by-using-the-same-kind-of-thinking-we-used-when-we-created-them-2012-4?r=US&IR=T
3. UNICEF (2014). *Syria's Children: A lost generation?* https://www. unicef.org/files/Syria_2yr_Report.pdf
4. Coulbeck, A., & Hugo, V. (1862). *Les Miserables.* Librairie internationale A. Lacroix

5. DQ Institute (2016). *2016 Singapore DQ Pilot Study*. https://www.dqinstitute.org/impact-research/
6. Sroufe LA, Carlson EA, Levy AK, & Egeland B. (1999). 'Implications of Attachment Theory for Developmental Psychopathology'. *Development and Psychopathology*, 11:1–13.

Chapter 7

1. DLSR – Working Group on Digital Literacy, Skills and Readiness (2020). *IEEE 3527.1-2020 – IEEE Approved Draft Standard for Digital Intelligence (DQ) – Framework for Digital Literacy, Skills and Readiness*. https://standards.ieee.org/standard/3527_1-2020.html
2. DQ Institute (2019). *DQ Global Standards Report 2019*. https://www.dqinstitute.org/wp-content/uploads/2019/03/DQGlobalStandardsReport2019.pdf
3. DQ Institute (2020). *2020 Child Online Safety Index: Findings and Methodology Report*. https://www.dqinstitute.org/wp-content/uploads/2020/02/2020-COSI-Findings-and-Methodology-Report.pdf
4. Sherlock, M., & Wagstaff, D. L. (2019). 'Exploring the relationship between frequency of Instagram use, exposure to idealized images, and psychological well-being in women'. *Psychology of Popular Media Culture*, 8(4), 482–490
5. Sang-Hun, C., & Lee, S (25 November 2019). 'Suicides by K-Pop Stars Prompt Soul-Searching in South Korea'. *The New York Times*
6. DQ Institute (2020). *Child Online Safety Index*. https://www.dqinstitute.org/child-online-safety-index/
7. DQ Institute (2020). *#DQEveryChild*. https://www.dqinstitute.org/dqeverychild/
8. American Academy of Paediatrics (2020). *Where We Stand: Screen Time*. https://healthychildren.org/English/family-life/Media/Pages/Where-We-Stand-TV-Viewing-Time.aspx
9. Park, Y. (27 January 2016). 'How much screen time should children have?'. *World Economic Forum*. https://www.weforum.org/agenda/2016/01/how-much-screen-time-should-children-have/
10. Dunkley, V.L. (14 February 2014). 'Gray Matters: Too Much Screen Time Damages the Brain'. *Psychology Today*. https://www.

psychologytoday.com/sg/blog/mental-wealth/201402/gray-matters-too-much-screen-time-damages-the-brain

11. Eisenberg, N., Hofer, C., & Vaughan, J. (2007). 'Effortful Control and Its Socioemotional Consequences' in J. J. Gross (ed.), *Handbook of Emotion Regulation*. The Guilford Press.

12. Mischel, W., & Ebbesen, E. B. (1970). 'Attention in delay of gratification'. *Journal of Personality and Social Psychology*, 16(2), 329.

13. Wachowski, L., & Wachowski, L. (1999). *The Matrix*. Warner Bros. Entertainment Inc.

14. John, A., Glendenning, A.C., Marchant, A., Montgomery, P., Stewart, A., Wood, A., Lloyd, K., & Hawton, K. (2018). 'Self-Harm, Suicidal Behaviours, and Cyberbullying in Children and Young People: Systematic Review'. *Journal of Medical Internet Research 20*(4): e129

15. IBM (2014). *IBM Security Services 2014 Cyber Security Intelligence Index*. https://www.ibm.com/developerworks/library/se-cyberindex2014/index.html

16. Harari, Y. (29 January 2015). *New Religions of the 21st Century*. Authors at Google.

17. Borba, M. (2016). *UnSelfie: Why Empathetic Kids Succeed in Our All-About-Me World*. Simon & Schuster.

18. Goleman, D. & Boyatzis, R.E. (6 February 2017). 'Emotional Intelligence Has 12 Elements. Which Do You Need to Work On?'. *Harvard Business Review*.

19. Bagchi, S. (2011). The Change Maker - Bill Drayton on Empathy and Leadership. *Forbes India*. https://www.forbesindia.com/article/zen-garden/the-change-maker-bill-drayton-on-empathy-and-leaderhip/25642/1

20. Forbes Quotes. *More Quotes by Lao-tzu*. https://www.forbes.com/quotes/5874/

21. Edwards, M. (14 April 2015). Identity theft: More than 770,000 Australians victims in past year. *ABC News*. http://www.abc.net.au/news/2015-04-14/identity-theft-hits-australians-veda/6390570

22. Jenkins Jr, H.W. (14 August 2010). Google and the Search for the Future. *Wall Street Journal*. https://www.wsj.com/articles/SB10001424052748704901104575423294099527212

23. Ronson, J. (12 February 2015). How One Stupid Tweet Blew Up Justine Sacco's Life. *The New York Times*. https://www.nytimes.

com/2015/02/15/magazine/how-one-stupid-tweet-ruined-justine-saccos-life.html

24. Longfield, A. (2017). 'Who Knows What About Me?'. *Children's Commissioner*. https://www.childrenscommissioner.gov.uk/digital/who-knows-what-about-me/

25. Ivanova, K. (18 August 2020). 'Terrified Mother Finds Stolen Images Of Her Baby Daughter On A Child Pornography Website'. *I Heart Intelligence*. https://iheartintelligence.com/terrified-mother-finds-stolen-images-of-her-baby-daughter-on-a-child-pornography-website/?fb=iis

26. May, A. (16 September 2016). '18-year-old sues parents for posting baby pictures on Facebook'. *USA Today*. https://eu.usatoday.com/story/news/nation-now/2016/09/16/18-year-old-sues-parents-posting-baby-pictures-facebook/90479402/

27. Vosoughi, S., Roy, D., & Aral, A. (2018) 'The Spread of True and False News Online'. *Science 359* (6380): 1146-1151.

28. Schlegel, L. (13 March 2020). 'Ready Player One: How Video Games Could Facilitate Radicalization Processes'. *European Eye on Radicalisation*. https://eeradicalization.com/ready-player-one-how-video-games-could-facilitate-radicalization-processes)

29. Meyer, R. (8 March 2018). 'The Grim Conclusions of the Largest-Ever Study of Fake News'. *The Atlantic*. https://www.theatlantic.com/technology/archive/2018/03/largest-study-ever-fake-news-mit-twitter/555104/

30. *Business Insider* (19 January 2014). 'Huge South Korean Data Leak Affects Almost Half The Country'. https://www.businessinsider.com/south-korea-data-leak-2014-1?r=US&IR=T

31. *Daily Mail* (6 February 2014). 'More than half of children use social media by the age of 10: Facebook is most popular site that youngsters join'. https://www.dailymail.co.uk/news/article-2552658/More-half-children-use-social-media-age-10-Facebook-popular-site-youngsters-join.html

32. World Economic Forum (2017). *Shaping the Future Implications of Digital Media for Society: Valuing Personal Data and Rebuilding Trust*. http://www3.weforum.org/docs/WEF_End_User_Perspective_on_Digital_Media_Survey_Summary_2017.pdf

33. Mckenna, M. (7 December 2020). 'A Conversation with GPT-3: Evaluating Issues Surrounding Bias, Ethics, Consent, and Overhype'. *Honey Suckle Mag.* https://honeysucklemag.com/a-conversation-with-gpt-3-bias/

34 Popkin, H.A.S. (13 January 2010). 'Privacy is dead on Facebook. Get over it.' *NBC News.* https://www.nbcnews.com/id/wbna34825225

Chapter 8

1. DQ Institute (2018). *2018 DQ Impact Report.* https://www.dqinstitute.org/2018dq_impact_report

2. DQ Institute (2020). *Child Online Safety Index.* https://www.dqinstitute.org/child-online-safety-index/

3. C20 Saudi Arabia (2020). *The C20.* https://civil-20.org/what-is-the-c20/

4. G20 Argentina (2018). *Toolkit for Measuring the Digital Economy.* *https://www.oecd.org/g20/summits/buenos-aires/g20-detf-toolkit.pdf*